Healthy and Productive Work

Healthy and Productive Work

An international perspective

Lawrence R. Murphy and
Cary L. Cooper

London and New York

First published 2000
by Taylor & Francis
11 New Fetter Lane, London EC4P 4EE

Simultaneously published in the USA and Canada
by Taylor & Francis Inc
325 Chestnut Street, 8th Floor, Philadelphia PA 19106

Taylor & Francis is an imprint of the Taylor & Francis Group

© 2000 Lawrence R. Murphy and Cary L. Cooper

Typeset in Times by
Florence Production Ltd, Stoodleigh, Devon
Printed and bound in Great Britain by
Biddles Ltd, Guildford and King's Lynn

British Library Cataloguing in Publication Data
A catalogue record for this book is available
from the British Library

Library of Congress Cataloging in Publication Data

ISBN 0–7484–0839–8

Contents

Figures and Tables

Figures

Tables

Contributors

Joyce A. Adkins, United States Air Force and Harvard Medical School, Department of Health Care Policy, 180 Longwood Avenue, Boston, MA 02115 (USA)

Guy Ahonen, PhD, Swedish School of Economics and Business Administration, PO Box 479, FIN–00101 Helsinki, Finland

Susan Cartwright, PhD, Manchester School of Management, UMIST, PO Box 88, Manchester M60 1QD, UK

Dr Cary L. Cooper, Manchester School of Management, UMIST, PO Box 88, Manchester M60 1QD, UK

Dianne Fassel, PhD, Newmeasures, Inc., 815 Laurel Avenue, Boulder, CO 80303 (USA)

Mark Griffin, PhD, School of Management, Queensland University of Technology, Brisbane 4001, Australia Email: M.Griffin@qut.edu.au

Peter Hart, PhD, Director, Social Research Consultants, University of Melbourne and Social Research Consultants, PO Box 712, Eltham Victoria 3095, Australia

R. J. L. Heron, AstraZeneca, Alderley House, Alderley Park, Macclesfield, Cheshire, SK10 4TF, UK

Simo Kaleva, MSc, Uusimaa Regional Institute of Occupational Health, Arinatie 3 A, FIN–00370 Helsinki, Finland

Kari Lindström, PhD, Finnish Institute of Occupational Health, Department of Psychology, Topeliuksenkatu 41 A, FIN–00250 Helsinki, Finland

Dr Noel McElearney, Marks & Spencer, London, UK

Karl O. Moe, Psychology Consultant to the Air Force Surgeon General, United States Air Force, Andrews AFB, MD (USA)

Jeff Monroy, Newmeasures, Inc., RR4, Box 207B, Ellsworth, Main 04605 (USA)

Margery Monroy, Newmeasures, Inc., RR4, Box 207B, Ellsworth, Main 04605 (USA)

Stacey Moran, PhD, St Paul Fire and Marine, Magnolia Court Suite F, 1716 Government Street, Ocean Spring, Mississippi 39564 (USA)

Lawrence R. Murphy, PhD, Senior Research Psychologist, National Institute for Occupational Safety and Health (NIOSH), 4676 Columbia Parkway, Cincinnati, Ohio 45226 (USA)

Timothy R. Parry, Vice President and General Council, Health Management Associates, Inc.

James Campbell Quick, PhD, Department of Management, Journal of Occupational Health Psychology, The University of Texas at Arlington, PO Box 19467, Arlington TX 76019–0467 (USA)

Wendi C. Schmidt, MEd, Manager of Corporate Health, The Home Depot, 2455 Paces Ferry Road, Atlanta, GA 30339

Kari Schrey, MA, Uusimaa Regional Institute of Occupational Health, Arinatie 3 A, FIN–00370 Helsinki, Finland

Eric Teasdale, PhD, Chief Medical Officer, Zeneca, Alderley House, Alderley Park, Macclesfield, Cheshire SK10 4TF, UK

John A. Tomenson, PhD, Dip. Stats. Cantab, BSc, Head, ICI Epidemiology Unit, PO Box 7, Brunner House, Winnington Northwich, Cheshire CW8 4DJ, UK

Linda Welch, MS, RD, LD, Wellness Supervisor, The Home Depot, 2455 Paces Ferry Road, Atlanta, GA 30339 (USA)

Lynne Christian Whatmore, MSc, Manchester School of Management, University of Manchester Institute of Science and Technology, PO Box 88, Manchester M60 1QD, UK

Steve Williams, PhD, Resource Assessment and Development, Claro Court, Claro Road, Harrogate HG1 4BA, UK

Mark G. Wilson, Department of Health Promotion and Behavior, 300 River Road, University of Georgia, Athens GA 30602–3422 (USA)

Ms Elisabeth Wilson-Evered, The University of Queensland and Queensland Health School of Psychology, The University of Queensland, St Lucia Queensland 4069, Australia

Preface

During the 1980s, "The Enterprise Culture" helped to transform economies in Western Europe and North America. But as we were to discover by the end of the decade, there was a substantial personal cost for many individual employees, both managers and shop floor workers. This cost was captured by a single word – *stress*.

Excessive pressure in the workplace was costly to business in the 1980s. For example, the collective cost of stress to US organizations for absenteeism, reduced productivity, compensation claims, health insurance, and direct medical expenses has been estimated at $150 billion per year. In the United Kingdom, stress-related absences were 10 times more costly than all other industrial relations disputes put together, and cost the UK economy £2 billion per annum.

Change will be the byword of the next millennium, with its accompanying job insecurities, corporate culture clashes, and significantly different styles of managerial leadership – in other words, massive organizational change and inevitable stress. In addition, change will bring with it an increased workload as companies try to become "leaner" to compete in the European, Far East, and other international economic arenas. This will mean fewer people performing more work, putting enormous pressure on them.

This book contains case studies of organizations that have introduced organizational health programs, from health screenings and stress management to employee involvement schemes. The book draws on good practice examples from the US, Europe, and Australia aimed at helping human resource managers explore what has been done and what the impact has been.

Cary L. Cooper
Lawrence R. Murphy

Chapter 1

Models of healthy work organizations

L. R. Murphy and C. L. Cooper

INTRODUCTION

The terms "healthy companies," "healthy work organizations," and "organizational health" all refer to the notion that worker well-being and
organizational effectiveness can be fostered by a common set of job
and organizational design characteristics. These represent significant depar-
tures from traditional models which have sought to improve either worker
health (e.g. health promotion) or organizational effectiveness (e.g. total
quality management). For the most part, these two lines of research inquiry
have been carried out independently and few empirical studies have sought
common antecedents or cross-cutting factors. Research on the topic of
healthy work organizations, on the other hand, seeks this common ground,
at the intersection of worker well-being and organizational effectiveness.
The aim of such research is to identify those job and organizational factors
which predict *both* health and performance outcomes.

The purpose of this chapter is to introduce the reader to the concept of
healthy work organizations in its various formulations, and to describe
theories or models which have been proposed to date on this topic. Since
articles on the topic of healthy work organizations have appeared in diverse
literatures, including human relations, health promotion, and job stress, a
secondary purpose of this chapter is to assemble this literature in one place
so that future research can build on a common knowledge base.

Although the term "healthy work organizations" has appeared only
recently in the organizational literature, the basic premise of improved
worker well-being plus organizational effectiveness is not new and has
been discussed in the organizational behavior, health promotion and job
stress literatures. Six models of healthy work organizations were identified in the research literature and are described in this chapter. As the
reader will see, the level of development of these models varies widely;
some models are very thorough, specifying antecedent, intervening, and
outcome variables. Other models are normative, and present only general
descriptions of major factors. It should be noted that none of the proposed

models has been rigorously tested across a wide range of occupations or industries; indeed, some of the models are based on data from a single company while others have not been empirically tested at all.

For ease of presentation, the models have been grouped into three sections corresponding to the field of study from which they originated: organizational behavior; health promotion; or job stress.

ORGANIZATIONAL BEHAVIOR

Perhaps more than any other area of study, the organizational behavior literature has produced more articles on the topic of healthy work organizations than any other field. Indeed, three of the six models reviewed in this chapter originated from this literature.

Goal integration model

In one of the earliest empirical works in this area, Barrett (1970) discussed various ways that social organizations could integrate the individual goals of health and well-being and organizational objectives of productivity and competitiveness. Based on the writings of human relations theorists (e.g. Argyris 1964; Herzberg 1966; Likert 1961; McGregor 1960), Barrett (1970) proposed three goal integration models: exchange, socialization, and accommodation. In the *exchange model*, workers exchange time and energy for incentives offered by the organization, i.e. they agree to work for pay and benefits. The classic organizational theory of bureaucracy aptly characterized such an economic exchange model. The *socialization model* achieves goal integration by social influence, encouraging workers to value those activities which lead to organizational objectives or devaluing activities which do not lead to the objectives. Persuasive communication and modeling is used to encourage workers to adopt or internalize organizational objectives or to abandon activities which interfere with the objectives. As examples, Barrett cited leader and peer socialization processes as discussed in the work of Schein (1967) and Likert (1961).

Finally, the *accommodation model* achieves integration by incorporating individual needs into the design of organizational objectives. The organization is structured in such a way that the pursuit of organizational objectives is intrinsically rewarding and will additionally lead to the attainment of individual goals. The writings of human relation theorists provide abundant examples of how job design, role design, and participation can foster the attainment of both worker needs and organizational objectives (e.g. Argyris 1964; Likert 1961; McGregor 1960).

Barrett (1970) tested the three goal integration models using questionnaire data obtained from 1781 employees of an oil refinery. The sample represented employees from most employment classifications and from all 18 departments. Multi-item scales were used to measure elements in each of the three goal integration models and indexes were formed to measure the degree of goal integration achieved, and organizational and team effectiveness.

The results provided strong support for the accommodation model, moderate support for the socialization model, and only limited support for the exchange model (i.e. for certain demographic subgroups). The accommodation model was the most stable across demographic factors, followed by the socialization and exchange models. Barrett (1970) concluded that the accommodation and socialization models were closely linked in practice and both were independent of the exchange model. The former models seemed to represent variations of a single management system which he described as a participative or democratic management.

Design features for a participative system

Drawing upon the same human relations literature as Barrett (1970), Lawler (1982) reached similar conclusions in his research on high involvement. He defined a "high involvement work organization" as one which fostered organizational effectiveness and employee quality of work life. He offered a detailed description of how to design an effective, high involvement work organization and constructed a model showing the key design variables, intervening variables, and well-being and performance outcomes.

Lawler's guidance on how to create an effective, high involvement work organization required attention to nine categories or systems:

1   *Organizational structure*: flat, lean, mini-enterprise oriented, team-based, participative structure.
2   *Job design*: individually enriched or self-managing teams.
3   *Information system*: open, inclusive, tied to jobs, decentralized-team based, participatively set goals.
4   *Career system*: career tracks and counseling available, open job posting.
5   *Selection*: realistic job preview, team-based, potential and process skill oriented.
6   *Training*: heavy commitment, peer training, economic education, inter-personal skills.
7   *Reward system*: open, skill based, gainsharing or ownership, flexible benefits, all salary, egalitarian perquisites.
8   *Personnel policies*: stability of employment, participatively established policies through representative group.
9   *Physical layout*: designed around the organizational structure, egalitarian, safe and pleasant.

In addition to providing great detail on job design, Lawler also specified two tiers in his model: one for organizational effectiveness and one for quality of work life. Within each tier, organizational characteristics were linked to outcomes via multiple intervening variables in a flow or path diagram.

Values-based organizational system

Rosen (1991) described the results of a three-year project to ". . . examine the intersection of human and economic concerns . . ." in US workplaces. Using a wide range of information sources which included newspapers, journals, books, company publications, plus hundreds of interviews with key business people, Rosen found that companies that fostered employee development and organizational effectiveness based their operations on eight core values. These values were: respect for all; leadership; managing change; life-long learning; workers are appreciating assets; sick jobs sabotage long-term investment; celebrating diversity; and work/family balance. He described the eight core values as ". . . the glue that binds healthy successful employees with healthy productive workplaces" and represented a set of perpetually interacting factors. Healthy companies demonstrate their commitment to the core values through actions, not just words. Thus, the values influence how workers treat each other, how managers communicate with workers, how jobs are designed, how reorganizations are undertaken, and how business decisions are made.

Based on these core values, Rosen developed a model of a healthy company which contained 13 dimensions:

- open communication
- employee involvement
- learning and renewal
- valued diversity
- institutional fairness
- equitable rewards and recognition
- economic security
- people-centered technology
- health-enhancing environment
- meaningful work
- family/work/life balance
- community responsibility
- environmental protection.

For an organization to be considered healthy, each dimension is required to be present both at the organizational level and the individual (worker) level. For example, the open communication dimension was operationally defined at the organization and individual levels by the statements: "The *organization* openly communicates about its operations and its plans and sharing occurs at all levels; *Individuals* respect the confidence of such information and participate in honest and forthright dialogue."

WORKSITE HEALTH PROMOTION

Health promotion involves education and motivational efforts to improve health and well-being through behavioral and lifestyle change. In work settings, these programs stand in contrast to traditional workplace health activities that seek to ensure worker protection from hazardous environmental conditions. Interest in health promotion programs has grown rapidly since the early 1980s, in large part because of soaring medical care costs and the realization that behavioral factors play a significant role in seven of the ten leading causes of death in the United States (Department of Health and Human Services 1979).

Corporate health promotion

Pfeiffer (1987) recommended that traditional health promotion programs be broadened from a singular focus on individual workers to include attention to team and organizational level factors. He proposed a model of corporate health that included three levels: *individual health*; *work team health*; and *organizational health. Individual health* is affected by heredity, the environment, lifestyle, and the medical care system, and interventions seek to improve the knowledge and decision-making of individual workers. Actions to improve individual health would include stress management, smoking cessation, exercise, nutrition, weight management, and hypertension screening. *Team health* focuses on the execution of assigned work, quality of services provided, nature of the work environment, and health and satisfaction of team members. Poor team health occurs when a team is forced to work short-handed, or when the team discourages individual participation in decision-making. Team health can be promoted by improved communication and problem-solving skills, conflict resolution training, peer support, employee involvement (e.g. quality circles), and occupational safety and health committees. *Organizational health* is a function of the interrelationship of the psychosocial work environment (i.e. the accepted or prescribed culture and norms), the quality of its products and services, the administrative systems that regulate day-to-day performance (e.g. policies, procedures, and program), and the employees themselves. Organizational health requires coordination of occupational health and safety, human resources, health promotion, medical services, and training/development functions. Actions to improve health at this level would include occupational health and safety programs, employee benefits, job security, compensation, smoking policies, health promotion programs, and educational assistance.

Pfeiffer (1987) proposed that a healthy work organization was one which offered "meaningful employment." Meaningful employment was defined as "a generalized set of values that the employee believes provides the satisfaction and rewards (fair compensation, job security, opportunity for advancement, safe/attractive work conditions, ability to make decisions) that make work self-fulfilling when compared with its sacrifices (low compensation, long hours, low autonomy)." Pfeiffer (1987) felt that the bottom line measure of meaningful employment was the "Personal Return on Investment (PROI)." The PROI considered the attributes of the work environment weighed against the downside of work. Low PROI was evidenced by poor quality work, low output, apathy, tardiness, excessive absenteeism, and poor morale.

In this same vein, Pelletier (1984) earlier had suggested that health promotion programs be expanded to include interventions aimed at improving environmental work conditions. In his book *Healthy People in Unhealthy Places*, he described the individual worker as the center of three concentric "spheres of influence," of which the team and organization were outer spheres. In this view, effective worksite programs to improve health would require attention to spheres beyond just the individual worker.

JOB STRESS

Over the past 20 years, there has been growing recognition of job stress as an important occupational health problem. In industry, there is a heightened awareness that occupational stress contributes to a significant portion of worker compensation claims, healthcare costs, disability, absenteeism, and productivity losses (Sauter *et al.* 1990). In a national survey by the Northwestern National Life, 46 percent of the 600 workers interviewed indicated that their job was very stressful, and 27 percent said it was the single greatest cause of stress in their lives (Northwestern National Life 1991).

In the sense that many models of job stress contain references to both worker health and performance outcomes, they could be considered healthy work organization models (e.g. Caplan *et al.* 1975; Cooper and Marshall 1976). On the other hand, they fall short of healthy work organization models because they usually separate health and performance outcomes, often testing them in separate, independent models.

Organizational health

In a series of editorials in the journal *Work and Stress*, Cox (Cox 1991; Cox and Cox 1992) recommended that job stress models be broadened to incorporate the concept of organizational health. He proposed that organizational health is affected by the consistency between the objective organization and the subjective organization. The objective organization refers to structure, policies and procedures. The subjective organization refers to the task of the organization, the way the organization perceives and solves problems, and the development environment provided to staff (i.e. employee growth). Unhealthy organizations are created when 1) there is inconsistency between the objective and subjective organization; 2) the subjective organization lacks coherence and/or is not well-integrated; or 3) the organization lacks a strong culture so that subsystems function rather independently, and often in conflict.

Cox and Cox (1992) proposed two examples of research questions which emerge from this new model: 1) To what extent does the health of the organization determine the health of individual worker or moderate the relationship between work and individual health? and 2) How does the health of the individual determine the structure, function, and culture of the organization? Research targeted to answer these questions presumably would generate a list of healthy work organization characteristics.

Healthy work organizations

Working in partnership with a private manufacturing company, the Finnish Institute of Occupational Health (FIOH) and the Manchester Institute for Science and Technology (UMIST), the National Institute for Occupational Safety and Health (NIOSH) developed an empirical model of a healthy work organization (Murphy and Lim 1997; Sauter *et al.* 1996). Three rounds of bi-annual employee climate survey were obtained from over 10,000 workers in 30 company locations and analyzed to empirically determine those characteristics associated with both worker well-being *and* organizational effectiveness.

The climate survey contained measures of *management practices* (e.g. leadership, strategic planning, employee performance rewards, career development); *organizational climate* (e.g. innovation, empowerment, diversity, intergroup cooperation); *corporate values* (e.g. individual worker, total quality, leadership, integrity); organizational performance (e.g. overall organizational effectiveness, workgroup performance, personal effectiveness); and worker well-being (e.g. job satisfaction, stress, turnover intent).

Canonical correlation analyses identified eight organizational characteristics associated with both organizational performance and worker well-being health outcomes: open, two-way communication, worker growth and development (training), trust and mutual respect, strong commitment to core values, strategic planning to keep the organization competitive and adaptive, rewards for performance, and workers being aware of how their work contributes to the business objectives (Lim and Murphy 1997).

Works not included

A number of works were not included in this chapter because they lacked an explicit link between worker health and organizational performance presented within a single model or they were not specific about key antecedent or outcome variables. These include the work of Peters and Waterman (*In Search of Excellence*, 1982), ecological models of health promotion offered by Stokols (1994), studies of successful companies by Collins and Porras (*Built to Last: Successful Habits of Visionary Companies*, 1994), and healthy work organization research at the Finnish Institute of Occupational Health (Lindstrom 1994) and the Swedish National Institute for Working Life (Aronsson 1996). Although they were not reviewed in this chapter, each of these works includes at least an implicit relationship between worker health and organizational performance as joint outcomes and so warrant attention by the reader.

In the same way, research on the relationships among employee satisfaction, organizational performance and customer satisfaction also deserves mention. While not proposing fully fledged models of organizational health, several researchers have demonstrated close interrelationships between good job design and organizational performance, and thus added to the empirical base needed for a more enlightened discussion of healthy work organizations (Ostroff 1992; Schneider and Bowen 1985; Turnow and Wiley 1991).

Conclusions

Information and data on characteristics of healthy work organizations are accumulating but theory continues to outrun scientific data. Although more empirical work is needed to test existing models and develop new ones, there appears to be a convergence of opinion and evidence on at least a few important factors. For example, most of the models described in this chapter emphasized meaningful work which utilizes worker skills, worker autonomy/control, job security, rewards for performance, worker involvement/participation and safe and healthy physical work environments as necessary components of a healthy work organization. There is a great deal of prior research which attests to the importance of these factors (e.g. Caplan *et al.* 1975; Herzberg 1966; Likert 1961; Sauter *et al.* 1990). Other factors noted in the more recent articles were conflict resolution/teamwork and commitment to core values. While these latter two factors have received less research attention than those mentioned above, they appear to be fruitful areas for additional research.

Despite this apparent convergence, a host of questions remain to be answered. For example, is there a generic healthy work organization model which is applicable to all industries or will some industries require specialized models? How does commitment to core values foster worker well-being and organizational health? What is the process by which core values become internalized (i.e. formalized) to the extent that they influence decision-making about purchasing, hiring, firing, and recruitment? In terms of measurement: Which objective measures of organizational performance and worker health/well-being are needed to supplement self-report data in future studies of healthy work organization models (e.g. return on equity, absenteeism, healthcare costs, disability)? In yet another vein: To what extent can external market conditions overpower or short-circuit the creation of healthy work organizations? How does shareholder concern for short-term vs. long-term profit influence the creation and maintenance of healthy work organizations?

Chapters in this book

The chapters in this book describe current efforts inside work organizations to create healthy work organizations. Each chapter has a dual focus – worker well-being *and* organizational performance – and most present data to empirically test the linkage between well-being and performance. As such, the chapters provide much needed evidence for the concept of healthy work organizations and should stimulate future research on this topic.

The scope of the chapters in this volume is noteworthy. Four deal with organizations in the US, three in the UK, one in Finland and one in Australia. With respect to the occupations and industries covered, three chapters deal with healthcare work, two with retail sales work, and one each with the military, metal and engineering industry, pharmaceuticals, and the government. In terms of interventions used to create healthy work organizations, four chapters address primary prevention efforts, and five describe secondary/tertiary prevention programs. Five of the chapters describe survey assessment and feedback procedures for identifying problem areas and designing targeted interventions, while the remaining four chapters discuss broad-based, company-wide policies and programs to foster employee health and organizational effectiveness.

By presenting efforts to create healthy work organizations from different countries, different occupation/industry settings, and varied intervention strategies, it is hoped that readers will find examples which suit their specific situation and can foster the process of designing healthy and productive workplaces.

References

Argyris, C. (1964) Integrating the individual and the organization, New York: Wiley and Sons.

Aronsson, G. (1996) Psychosocial issues at work: current situation and trends in Sweden, in *Proceedings of Occupational Health and Safety in Progress:*

Northern-Baltic-Karelian Regional Symposium, Lappeenranta, Finland: Finnish Institute of Occupational Health.

Barrett, J. (1970) *Individual Goals and Organizational Objectives: A Study of Integration Mechanisms*, Ann Arbor, Michigan: Center for Research on Utilization of Scientific Knowledge.

Caplan, R. D., Cobb, S., French, J. R. P. Jr, Harrison, R. V. and Pinneau, S. R. (1975) *Job Demands and Worker Health*, DHHS (NIOSH) Publication No. 75-160, Washington, DC: US Government Printing Office.

Collins, J. and Porras, J. (1994) *Built To Last: Successful Habits of Visionary Companies*, New York: HarperBusiness.

Cooper, C. L. and Marshall, J. (1976) Occupational sources of stress: a review of the literature relating to coronary heart disease and mental ill health, *Journal of Occupational Psychology* 49: 11–28.

Cox, T. (1991) Organizational culture, stress, and stress management, *Work and Stress*, 5: 1–4.

Cox, T. and Cox, S. (1992) Occupational health: past, present and future, *Work and Stress* 6: 99–102.

Department of Health and Human Services (1979) *Healthy People: The Surgeon General's Report on Health Promotion and Disease Prevention*, DHHS (PHS) Publication No. 79-55071, Washington, DC: US Government Printing Office.

Herzberg, F. (1966) *Work and the Nature of Man*, Cleveland: World Publishing Company.

Lawler, E. E. (1982) Increasing worker involvement to enhance organizational effectiveness, in P. Goodman and Associates (eds) *Change in Organizations*, pp. 280–315, San Francisco: Jossey-Bass.

Likert, R. (1961) *New Patterns of Management*, New York: McGraw-Hill.

Lim, S. Y. and Murphy L. R. (1997) Models of healthy work organizations, in P. Seppälä, T. Luopajärvi, C-H. Nygård and M. Mattila (eds) *From Experience to Innovation (Volume 1)*, Helsinki, Finland: Finnish Institute of Occupational Health.

Lindstrom, K. (1994) Psychosocial criteria for good work organization, *Scandanavian Journal of Work Environment Health* 20 (special issue): 123–33.

McGregor, D. (1960) *The Human Side of Enterprise*, New York: McGraw-Hill.

Murphy, L. R. and Lim, S. Y. (1997) Characteristics of healthy work organizations, in P. Seppälä, T. Luopajärvi, C-H. Nygård and M. Mattila (eds) *From Experience to Innovation (Volume 1)*, pp. 513–15, Helsinki, Finland: Finnish Institute of Occupational Health.

Northwestern National Life Insurance Company (1991) *Employee Burnout: America's Newest Epidemic*, Minneapolis, MN: Northwestern National Life Insurance Company.

Ostroff, C. (1992) The relationship between satisfaction, attitudes, and performance: An organizational level analysis, *Journal of Applied Psychology* 77: 963–74.

Pelletier, K. R. (1984) *Healthy People in Unhealthy Places. Stress and Fitness at Work*, A Merloyd Lawrence Book, Delacorte Press/Seymour Lawrence.

Pfeiffer, G. J. (1987) Corporate health can improve if firms take organizational approach, *Occupational Health and Safety*, October: 96–9.

Rosen, R. H. (1991) *The Healthy Company. Eight Strategies to Develop People, Productivity, and Profits*, Los Angeles: Jeremy P. Tarcher, Inc.

Sauter, S. L., Lim, S. Y. and Murphy, L. R. (1996) Organizational health: A new paradigm for occupational stress research at NIOSH, *Japanese Journal of Occupational Mental Health*, 4: 248–54.

Sauter, S. L., Murphy, L. R. and Hurrell, J. J. Jr (1990) A national strategy for the prevention of work-related psychological disorders, *American Psychologist*, 45: 1146–58.

Schein, E. (1967) *Organizational Socialization and the Profession of Management*, Cambridge, Massachusetts: MIT Press.

Schneider, B. and Bowen, D. (1985) Employee and customer perceptions of service in banks: Replication and extension, *Administrative Science Quarterly* 70: 423–33.

Turnow, W. W. and Wiley, J. W. (1991) Service quality and management practices: A look at employee attitudes, customer satisfaction, and bottom-line consequences, *Human Resource Planning* 14: 105–15.

Part I

Targeted organizational change based on employee survey data

Chapters in this section approached the task of creating healthy work organizations by first conducting an employee survey to identify those organizational characteristics associated with employee good health and performance. Based on the survey results, specific, targeted interventions were designed and implemented. The first three chapters deal with healthcare organizations. With a distinct focus on testing a model of organizational climate, Griffin, Hart, and Wilson-Evered describe an assessment and intervention in a large urban hospital in Australia. The authors provide a detailed background on conceptual issues surrounding program development, the role of survey feedback in the intervention process, and the value of internal and external benchmarking. A range of organizational change interventions were implemented (e.g. improved communication, team recognition, professional development) and their effects on employee well-being assessed.

Fassel, Monroy, and Monroy describe a unique approach to organizational change in a large US healthcare system which has as its centerpiece the integration of core values into the work experience. Using survey data obtained from over 20,000 healthcare workers across 50 different hospitals, path analytic models were tested and the key drivers for improving value integration and organizational effectiveness identified in the analyses were used to develop organizational change interventions.

The chapter by Moran and Parry emphasizes the importance of adapting to change in the turbulent healthcare world of hospital acquisitions. Their study of over 7000 workers in 26 recently acquired healthcare organizations highlighted the importance of effectively managing the human-side of change by showing that employee perceptions of managing change were associated with patient satisfaction and employee turnover.

Cartwright, Cooper, and Whatmore focus on communication as a key element of organizational culture. Using the results of a stress audit conducted in a UK government agency which employs 25,000 workers, the authors worked with the agency to design an intervention program which involved a number of initiatives to improve communication.

Post-intervention assessment of the intervention revealed improvements in job satisfaction, and employee perceptions of control and influence.

The final chapter in this section is the most comprehensive, in terms of the occupations/industries studied and the type of intervention tested. Lindström, Schrey, Ahonen, and Kaleva collected data on organizational and employee reactions from over 4000 Finnish workers in 281 workplaces representing 12 difference industries. The interventions varied widely across study sites but included improved management practices, communication, and multi-skilling. Positive benefits of the interventions to workers and the organization are described.

Chapter 2

Using employee opinion surveys to improve organizational health[1]

*M. A. Griffin, P. M. Hart, and
E. Wilson-Evered*

ORGANIZATIONAL HEALTH

Aligning employee well-being and organizational effectiveness is one of the key goals of interventions designed to improve organizational health. The organizational health perspective differs from many of the traditional approaches to reducing occupational stress in two important ways. First, it emphasizes the need to focus on *both* employee well-being and the organization's "bottom-line" performance (Cox 1988; Miller *et al.* 1999). This perspective recognizes the fact that having happy and satisfied employees is of little value to an organization unless the employees are performing efficiently and productively. Likewise, having an efficient and productive organization is of little value if this is achieved at the expense of employees' well-being. Although this view is intuitively appealing, little is known about how we can simultaneously improve employee well-being and organizational performance.

Second, the organizational health perspective recognizes that employee well-being and organizational performance are determined by both individual *and* organizational factors. This view is consistent with recent developments in the occupational stress literature. For example, the dynamic equilibrium theory of stress proposed by Hart (in press) suggests that stress results from a broad system of variables that includes personality (e.g. Barrick and Mount 1991; Costa and McCrae 1980) and organizational (Michela *et al.* 1995) characteristics, coping processes (e.g. Bolger 1990), positive and negative work experiences (e.g. Hart *et al.* 1995b), and various indices of psychological well-being (e.g. George 1996). As noted by Lazarus (1990), however, stress cannot be located in any one of these variables. Rather, stress only occurs when a state of disequilibrium exists within the system of variables relating people to their environments, provided that this state of disequilibrium brings about change in people's normal (i.e. equilibrium) levels of psychological well-being. Although the emphasis on stability and change distinguishes the dynamic equilibrium theory from other theoretical perspectives, the theory recognizes that stress results from the dynamic interplay between employees and their work environments. This is consistent with many other approaches to occupational stress (e.g. Edwards 1992; Lazarus and Folkman 1984), and is shown diagrammatically in Figure 2.1.

The role of employee opinion surveys

In our experience, one of the most effective ways of improving organizational health is to involve an organization's employees in a comprehensive and systematic assessment of their current operating practices, and to link this assessment to programs of continuous improvement that have a focus on employee well-being and performance. This can be achieved through the conduct of a well-designed employee opinion survey (Kraut 1996). To be effective, however, the employee opinion survey must be integrated into the organization's operating procedures (e.g. business planning, human resource management practices, and performance management), as well as being linked to the review processes and decision-making at all levels within the organization. When this occurs, an employee opinion survey can be a powerful tool for communicating information about employee well-being and organizational effectiveness.

In this chapter we draw on a hospital-based intervention program to illustrate how employee opinion surveys can be used to improve organizational health. We begin by discussing the central role that organizational climate plays in determining organizational health. We then describe the environment of change that the hospital was working in, and outline the survey methodology and organizational activities that were implemented in response to the survey process. The diagnostic and evaluative components of the survey are illustrated by drawing on selected analyses

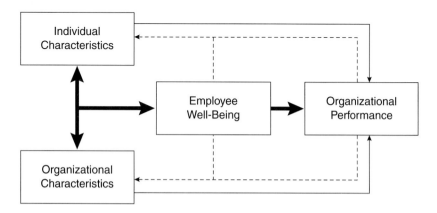

Figure 2.1   Organizational health flow diagram

that were conducted on the two waves of longitudinal data collected from the hospital's employees. In addition to the employee opinion data, the analyses used information obtained from staff and management about the success of the intervention strategies that were designed to improve organizational health.

THE CENTRAL ROLE OF ORGANIZATIONAL CLIMATE

We believe that a positive organizational climate is a critical indicator of organizational health. The climate of the organization is the sum of the processes and activities within the organization *as they are perceived by the organization's staff.* This definition highlights that both organizational activities and staff perceptions of those activities determine the climate of any organization. Although this means that organizational climate is largely subjective in nature, it also means that it has a component that is shared by all employees. In other words, it is both an individual and an organizational level construct.

What is organizational climate?

Organizational climate refers to the perceptions that staff have about the way in which their workplace functions (James and McIntyre 1996). These perceptions concern policies, practices, and rewards at work (Neal and Griffin 1998; Schneider 1990). Climate can also describe specific areas of work such as safety climate (Neal and Griffin 1998) and climate for customer service (Jimmieson and Griffin 1998). In each case, climate is about employee perceptions of how the workplace operates.

A wide range of studies has shown that staff can make meaningful judgements about the quality of their work environment. Staff are able to clearly distinguish between the different features of organizational climate, and often reach high levels of agreement about the way in which their organization functions. This has been a consistent finding in our research and consulting work across a range of private and public sector organizations (e.g. Hart and Wearing 1999; Hart *et al.* 1997). Moreover, this work has led to the identification of 10 key features of the work environment that are relevant to all organizations (Hart *et al.* 1999), and it was these features that formed the basis of the hospital's employee opinion survey:

- the quality of leadership (*supportive leadership*)
- the decision-making processes that are used in the workplace (*participative decision-making*)
- the clarity of staff roles (*role clarity*)

- the way in which staff support one another and work together (*professional interaction*)
- the extent to which staff receive appraisal and recognition (*appraisal and recognition*)
- opportunities for professional growth (*professional development*)
- the extent to which staff agree with the goals and philosophy of the workplace (*goal congruence*)
- the level of work demands (*excessive work demands*)
- the overall level of enthusiasm, energy and pride in the workplace (*workplace morale*)
- the overall level of anxiety, distress and frustration in the workplace (*workplace distress*)

These 10 different aspects of organizational climate can be applied across a range of different industry types (e.g. education, manufacturing, police, telecommunications) and occupational groups (e.g. clerical, professional, trades) (Hart *et al.* 1997). Moreover, our research and consulting work suggests that these organizational characteristics are among the critical factors that determine staff well-being and bottom-line outcomes, such as absenteeism, retention rates, individual and group performance, service delivery and the reputation of the organization amongst its customer base (Hart *et al.* 1995a, b, 1996b).

This doesn't mean that different occupations do similar work. Clearly, the work of a teacher is quite different from that of a police officer, which is different from that of an accountant or an airline pilot. Although the operational work of various occupational or industry groups may differ, the organizational aspects of most jobs tend to be quite similar (e.g. involving leaders, requiring role clarity, involving appraisal and recognition process, etc.). Moreover, research has shown that it is the organizational rather than the operation aspects of a job that tend to be most stressful, even among occupational groups such as police officers (Hart *et al.* 1995b) and teachers (Hart, 1994, 1995a).

Differentiating organizational climate from other perceptions

Although organizational climate is based on employee perceptions of their work environment, not all employee perceptions are about organizational climate. Perceptions of the work environment (organizational climate) can be clearly differentiated from other types of perception. For example, it is possible to make clear distinctions between employees' perceptions about:

- specific feelings or judgements associated with work (e.g. anxiety, satisfaction, distress)

- performance-related behaviors at work (e.g. doing a good job, participating in meetings, team work)
- attachment behaviors (e.g. absenteeism, turnover)
- individual personality (e.g. typical levels of conscientiousness and enthusiasm)
- organizational outcomes (e.g. reputation, customer service, performance)

Perceptions that are related to one or more of these factors are often assessed without making distinctions among the different types of measures that should be used during the assessment. This can be problematic for two reasons. First, it is important to accurately distinguish between different types of perception so that it is possible to demonstrate discriminant validity among the different factors that are assessed in an employee opinion survey. Second, and more importantly, it is necessary to demonstrate discriminant validity before the data obtained during an employee opinion survey can be used to benchmark organizational health and establish how different determinants of organizational health (e.g. climate) contribute to employee well-being and performance. These reasons highlight the need to ensure that employee opinion surveys are based on high quality measures, in terms of their reliability *and* validity, if the resulting data are to be used for diagnosing and monitoring organizational health.

Diagnosing organizational climate

Although many organizational climate surveys have good face validity (i.e. they appear to ask the right questions), this is not a sufficient test of validity if the resulting information is to be used to develop and implement effective intervention strategies. As demonstrated by Hart and Wearing (1999), items with good face validity may still be unreliable, or may fail to measure the feature of organizational climate they are thought to assess. When this occurs, the information obtained from an employee opinion survey is of little diagnostic value. If it can be demonstrated that the organizational climate survey has good reliability and validity, however, it is then possible to use the resulting information in one of two ways: 1) *benchmarking* to assess the levels of organizational health; and 2) *mathematical modeling* to assess the key drivers (i.e. "causes" or determinants) of organizational health.

Benchmarking

The benchmark comparisons that can be obtained from an employee opinion survey provide important information for individual workgroups and for the organization as a whole. Two main types of benchmark comparisons can be made; *external* and *internal*. These comparisons each provide different types of information about how well an organization is performing. This information can be used in a variety of ways to inform decision-making and the development of strategies aimed at improving organizational health. The key strengths of benchmark comparisons are that they:

- summarize the attitudes that staff have on a wide range of issues
- succinctly communicate information about the performance of a workplace
- provide a structure for identifying critical issues, sources of concern, and areas of success in the workplace; and
- provide a reference point for critical discussion about organizational health within the workplace.

External benchmarks compare the results of an employee opinion survey for the organization as a whole to the results obtained from staff working in other organizations. This is a useful comparison if an organization wants to establish how well it is performing in relation to other organizations. The extent to which the comparison provides meaningful information, however, depends on whether the comparative organizations are similar in some important way (e.g. they belong to the same industry sector or staff perform similar types of jobs).

One of the limitations of external benchmark comparisons, however, is that they do not, in themselves, indicate whether the level of organizational health is good, bad or indifferent in a particular organization. For example, an external benchmark comparison may indicate that the organization is performing at a level that is 10 percent better than similar organizations. Although, at face value, this suggests that the organization is performing well, this type of comparison does not tell us whether or not this level of performance is a good outcome for the organization in question. If the performance of the organization has been improving over time, then this would be a good outcome. If, however, the organization were performing at a lower standard than, say, 12 months earlier, then this would be a poor outcome, even though the benchmark level is still higher than that obtained in similar organizations. In other words, the performance of an organization may depend more on its trendline over a period of time, rather than how well it may be doing at any one point in time.

A second limitation of external benchmark comparisons is that they do not provide information about the strengths ("success stories") and weaknesses ("hot spots") within an organization. For example, an external benchmark comparison may suggest that an organization is performing well in comparison to similar organizations. This may mask the fact, however, that there are many workgroups within the organization that are performing quite poorly. In other words, the average benchmark score for an organization tells us nothing about the variation in scores within the organization. Our research and consulting work has shown that there is substantially more variation among the workgroups within an organization than there is between organizations. This variation can go unnoticed when relying solely on external benchmark comparisons.

Internal benchmarks compare the results for discrete workgroups (e.g. departments, sections or work teams) with other results that have been obtained within the same organization. This can take a number of different forms. For example, it is possible to compare the results for discrete workgroups with the results that were obtained for the organization as a whole. This can be useful in helping to identify the high and low performing workgroups within the organization. One of the limitations of this type of comparison, however, is that by design, approximately half of the workgroups will be performing below average, whereas the other half of the workgroups will be performing above average.

It is also possible to compare the performance of a particular workgroup with itself over time. In our experience, this type of benchmark comparison can be very effective in motivating improved organizational health. Moreover, it enables a workgroup to monitor the change in its own level of performance. One of the limitations of this type of comparison, however, is that it may lead a workgroup to focus on itself, having little regard for how it is performing relative to other workgroups and the organization as a whole.

There is no single benchmark comparison that is ideal for diagnosing organizational health. Rather, each comparison provides a different type of information that can be used to determine how well an organization is performing in terms of its levels of organizational health. This means that a range of benchmark comparisons should be used to establish the levels of organizational health within an organization. It is important, however, to note that benchmark comparisons merely provide static or structural information about the state of an organization (Hart and Wearing 1999). In other words, benchmark comparisons do not provide dynamic information that describes how the organization functions.

Modeling the causes and consequences of climate

To formulate policy and develop strategies for improving organizational health, it is necessary to have information about the cause and effect relationships among the factors that contribute to organizational health. In other words, it is important to obtain information about the key drivers that contribute to employee well-being and organizational performance.

Establishing cause and effect relationships is not straightforward, particularly in a non-experimental setting, such as an organization. As noted by Hart and Wearing (1999), however, it is possible to apply mathematical modeling techniques, such as structural equation analyses (e.g. Cuttance and Ecob 1987), to employee opinion data to:

- explicitly test competing causal hypotheses
- control for the effects of extraneous factors
- facilitate the forecasting and cost benefits analyses that should guide organizational change
- provide a means for understanding the complexity of change and its consequences

Although it is preferable to apply these techniques to employee opinion data that have been collected from the same employees on at least two different occasions, it is also possible to use the results of a single survey to gain some insight into the factors that contribute to organizational health. For example, we have used the information obtained from single employee opinions surveys to examine whether employee well-being is determined more by the climate of an organization, the personalities of employees, the coping strategies that employees use to deal with day-to-day difficulties, or employees' positive and negative work experiences. By employing mathematical modeling techniques to the employee opinion data, we have found that organizational climate is the primary determinant of employee well-being in a range of organizations (e.g. Hart et al. 1996a, b), including the organization that was used for the case study reported in this chapter. These results have highlighted the need to focus on improving the climate of an organization in order to maximize its levels of organizational health.

Although there are many limitations that must be taken into account when using mathematical modeling techniques with employee opinion data, a careful application of these techniques can provide valuable diagnostic information that is not available through the use of more traditional benchmark comparison analyses. Moreover, it is possible to supplement the information obtained from employee opinion surveys with other sources of data, such as absenteeism rates, retention rates, information on key performance indicators, customer or client satisfaction ratings, and other third party observations about the performance of individual employees and their workgroups. By integrating different sources of data, it is possible to increase the reliability and validity of mathematical modeling procedures, and add greater diagnostic value to an employee opinion survey.

ORGANIZATIONAL CASE STUDY

The organizational context

An urban women's hospital

A project designed to improve organizational health was conducted in an Australian urban hospital that employs approximately 1000 staff. The hospital is publicly funded and its primary clinical services include obstetrics, gynaecology, gynaecological-oncology and neonatology. Recent changes in the population demographics, epidemiology, and approaches to clinical care have all impacted upon the community's demands for these specialized healthcare services. Furthermore, worldwide growth and development in the healthcare industry, together with shifts in government funding and service priorities, have created an environment of continuous change at the hospital.

In addition to the external forces creating an environment of change, there were also a number of internally driven changes involving redevelopment and restructuring. Restructuring has included changes in senior management, some closure of wards and reduction in staffing levels, and an emphasis on business planning. These changes aim to improve the overall performance of the hospital and were designed to bring about benefits for patients, staff and other key stakeholders.

The organizational improvement initiative

As part of the project, all hospital staff were invited to participate in an organization-wide employee opinion survey in 1996 and again in 1997. The surveys were designed so that the hospital could 1) conduct an organization health audit (diagnosis); 2) develop and implement a range of initiatives to address the issues identified through the organizational health audit; and 3) evaluate the effectiveness of these initiatives. The survey was based on a questionnaire that had been used in a wide variety of public (Hart et al. 1996a) and private (Hart et al. 1997) sector organizations. The questionnaire assessed a range of organizational health indicators, including 10 different dimensions of organizational climate, individual distress and morale (also known as work-related positive and negative affect), turnover intentions, and number of sick days taken without a medical certificate (non-certified sick leave).

A range of resources and activities was implemented to support the survey process. A project officer was appointed and an external consultant was contracted for survey development, analysis, and support in feedback dissemination. Senior executive support for the project was demonstrated through letters and memos to staff and managers. A communication strategy was implemented that included briefings at key staff meetings, a video about the project, individual interviews with managers, information sheets, posters, and newsletters. An overall response rate of 56 percent was achieved.

Feedback about the survey results was provided to the project steering committee, the hospitals senior management group, all functional work-groups, and representatives of all professional groups. A report from the first survey was produced through collaboration of the consultant, project officer, and hospital representatives. The report was published in March 1997 and distributed to each departmental and line-manager throughout the organization. After extensive consultation, an Intervention Plan was produced in November 1997 that described strategies to improve organi-zational health. These strategies focused on the priorities identified from the results of the employee opinion survey. The Intervention Plan also contained a report that was based on interviews with each workgroup that aimed to identify team level initiatives.

The overall goal of the Intervention Plan, in conjunction with the continual review of other performance data, was to develop a climate of continued learning and improvement. Workgroup and expert ratings about the effectiveness of the organizational health improvement initia-tives were collated in 1997 and 1998. Other performance data, such as the sick leave rates, turnover rates, and incident or accident rates for each workgroup, were also collated. Through the processes of data acqui-sition, analyses and feedback, workgroups became increasingly interested in developing an understanding of the elements of a healthy organization or workgroup.

In summary, the organization-wide surveys provided a means of assessing 1) employee well-being and satisfaction; 2) the perceived effec-tiveness of the change initiatives; and 3) the change, for better or worse, that occurred on a range of organizational health indicators during the course of the project. By focusing on natural work or professional groups, the results of the surveys were used to encourage group level responsi-bility for examining their current situation and designing strategies that would bring about a positive change in the levels of organizational health. In this way, the surveys facilitated collaborative interventions throughout all levels of the organization.

The initial diagnosis

Benchmarks – a workgroup analysis

An example of the benchmark results for the first year of the survey is presented in Figure 2.2. This figure shows the results that were obtained on the key organizational health indicators for an individual work unit

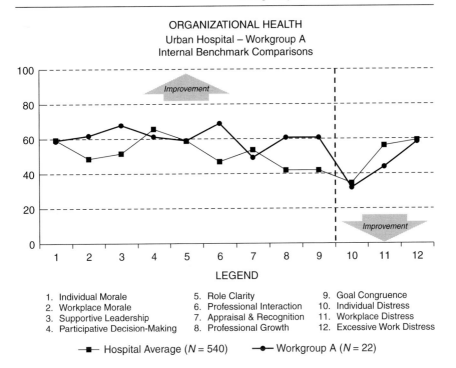

ORGANIZATIONAL HEALTH
Urban Hospital – Workgroup A
Internal Benchmark Comparisons

LEGEND

1. Individual Morale
2. Workplace Morale
3. Supportive Leadership
4. Participative Decision-Making

5. Role Clarity
6. Professional Interaction
7. Appraisal & Recognition
8. Professional Growth

9. Goal Congruence
10. Individual Distress
11. Workplace Distress
12. Excessive Work Distress

Hospital Average ($N = 540$)   Workgroup A ($N = 22$)

Figure 2.2   Internal benchmark comparison graph showing the results on indicators of organizational climate and occupational well-being for Workgroup A in comparison to the average result for the hospital as a whole

in comparison to the results for the organization as a whole. This type of benchmark graph was provided to all functional workgroups and professional groups in the hospital as part of the initial feedback on the survey. Each workgroup was also provided with supporting documentation that described how to interpret their graph, as well as resources for developing responses to the survey information.

Overall, the hospital level benchmarks indicated that the organizational climate was somewhat less positive than comparison organizations in the same state. The workgroup results indicated a large degree of variation in the climate dimensions. Some groups reported consistently positive results across all climate dimensions; other groups reported consistently negative results, and some groups reported wide differences in the levels of each climate dimension. The variation that was observed among workgroups demonstrates that internal benchmarks provide important diagnostic about the strengths ("success stories") and weaknesses ("hot spots") within an organization.

Modeling organizational climate – an organizational analysis

Mathematical modeling of the climate dimensions was an important source of information for developing organization-wide improvement strategies. Figure 2.3 presents the mathematical model that was used to inform organizational initiatives. The model provides a comprehensive description of the way organizational climate influences both employee well-being and outcomes such as uncertified sick leave and turnover. The arrows in the model show the paths of influence among the measures of the survey. Numbers represent the strength of the path and can range from −100 to +100. Values close to zero describe weaker relationships than paths with higher values. A path with a positive value means that an increase in the dimension at the tail of the arrow will result in an increase in the dimension at the head of the arrow. A path with a negative value means increase in the first dimension will result in a decrease in the second dimension.

The arrows in the model depict direct relationships among the survey dimensions. When there is no arrow between any two dimensions, it may still be possible to identify a path of influence. For example, there is no arrow linking the dimension of Appraisal and Recognition and the dimension of Individual Morale: however, there is an indirect link via Professional Growth. This indirect link is very important because it explains how the climate dimension of Appraisal and Recognition actually functions to improve well-being. Therefore, a detailed analysis of the model can provide a very informative picture of the way that climate operates to influence outcomes in the workplace.

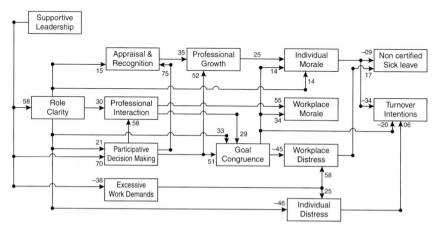

Figure 2.3   Mathematical model linking climate dimensions, well-being indicators, and outcomes

An effective way to summarize the information contained in the model is to look at the overall impact of the direct and indirect paths on well-being and other outcomes. Table 2.1 shows the results of this summary. The table lists, in rank order, the most important influences on workplace morale and distress and on turnover intentions and sick leave. Together, the model and the summary table provide a detailed picture of how climate drives important organizational outcomes.

The information from the model and summary table was used by the steering committee, consultants, and management groups to understand the impact of climate and to identify priority areas for improving organizational health. A key outcome of this process was recognition that leadership played an important role in all of the outcomes. The results indicated that supportive leadership was a pervasive influence on employees' feelings about their work and their intentions for staying in the organization.

The process of analysis and interpretation undertaken in the hospital resulted in a range of conclusions used to develop action plans for improvement. For example, participative decision-making was directly and strongly related to appraisal and recognition. Although leadership was strongly related to appraisal and recognition, the impact of leadership was indirect and transmitted largely via participative decision-making and role clarity. This finding suggests that leaders influence the degree individuals feel valued by increasing the degree to which they engage them in defining their roles and making decisions about issues that affect them in the workplace. The model indicates that appraisal and recognition is also directly related to individual morale and, in turn, to sick leave and turnover intentions. Therefore, a key finding is the link between supportive leadership, participative decision-making and appraisal and recognition.

A second implication of the modeling was the distinction between individual well-being measures of distress and morale and the overall measures of workplace distress and morale. The individual and the workplace measures had different antecedents and different influences on the outcomes. For example, professional interaction was directly related to workplace morale but not individual morale. This result indicates that it is important to consider a range of outcomes when designing organizational initiatives to improve organizational health.

Table 2.1 Strongest influences from climate model on four outcomes

Workplace morale	Workplace distress	Turnover intentions	Sick leave
Leadership	Workload	Leadership	Workplace distress
Professional interaction	Leadership	Role clarity	Leadership
Decision-making	Goal congruence	Goal congruence	Workload

A final implication for organizational health was that bottom-line outcomes were driven by both negative and positive dimensions of well-being. Traditional approaches to work stress focus on the negative aspects of well-being. The hospital results demonstrated that the positive dimensions of well-being (individual morale and workplace morale) have distinct causes and have an independent impact on organizational health outcomes.

For example, work demands increased distress at work but did not decrease feelings of morale. Professional growth, on the other hand, increased morale but did not decrease distress. In addition, both morale and distress influenced the key outcome of turnover intentions. Therefore, focusing only on the negative aspects of well-being would show a very incomplete picture of the way that climate influences organizational health. These results also indicate that interventions that aim to increase positive feeling of staff need to go beyond workload and address other issues of organizational climate such as leadership and recognition in the workplace.

In summary, the results highlighted that a variety of climate factors were important for determining employee outcomes. As a consequence, there is no one simple solution for improving organizational health. A key theme that emerged from the analysis was that participation in decision-making was important for a range of employee outcomes. For the hospital, it appeared particularly important for supervisors and leaders to encourage processes that allow staff to have a say in their workplace.

Using the survey results

The survey results were integrated in various ways in the organization. At the workgroup level, local work teams made improvements based on the findings of the survey specific to their workgroup. Professional groups developed action plans that would contribute to increased morale levels across the profession. Organization level plans were developed based on priorities identified using the external benchmarks and mathematical models.

Organizational improvement initiative teams

The results provided data for a range of strategies, which were implemented over the year. Two teams (Organizational Improvement Initiative Implementation Team and Workplace Improvement Team) and several subgroups developed strategies to improve staff well-being and satisfaction. Examples include a communication strategy, team identification and recognition, assistant district manager's forum, access to professional development and courses aimed at improving both work skills and interpersonal relations and the leadership development program. The hospital accreditation process enabled across-discipline teamwork that encouraged professional interaction – a key factor recognized as contributing significantly to staff morale in the hospital. Although there have been significant changes in leadership, employees at all levels continued to focus energies on patient care and customer satisfaction.

Leadership

For their part the senior leadership set the strategic direction, became more visible and involved staff in problem solving and decision-making on issues relevant to the workplace. A new clinical divisional structure is expected to provide further enhancement through the devolution of organizational decision-making to the division. The impact will be seen especially with respect to staff, clinical services and budget decisions which will be made locally. The staff satisfaction results have been reported in the District Improvement Plan and the initiative was commended by the team of surveyors who recommended the accreditation of the hospital. Further publication and integration of the staff satisfaction results and appraisals of the organizational climate are expected in the future.

Work or professional areas improvement

Some work areas responded to the survey with interest and enthusiasm and developed focused action plans to address areas of concern and maintain areas of strength. The early development of action plans in work areas produced outstanding effects on staff relations within a very short time. Other groups worked consistently on achieving improvements on complex issues throughout the year. Staff in these areas shared responsibility with the workgroup leader or line manager in the design, implementation and evaluation of the improvement strategy. The work environment improvement focus uses the same approach as already applied to other aspects of work such as service delivery (identify issues, obtain data, plan change, implement, evaluate, review).

Examples of workgroup improvements based on the survey findings are:

- the deliberate structuring of multi-disciplinary team meetings
- developing improved methods of communicating and sharing information across shifts and groups
- staff mentoring and peer support program
- staff development, education and training
- staff presentations, seminars and in-service programs to address identified learning needs related to work roles and team performance
- development and encouraging an environment of learning and growth
- planning with a range of groups to improve patient-care processes

- developing local means to recognize achievement
- collaborative goal setting and associated projects to achieve goals
- updating role definitions with appropriate skills enhancement
- open-door to manager policy

Each workgroup had the opportunity to reflect on their survey findings and make collective interpretations and collaborative plans to suit their requirements. Groups shared their efforts and achievements with other groups at a range of venues.

Framework for improvement

This organization put resources into the critical analysis of employee perceptions of the organizational environment. The hospital has been at the forefront of this movement in the state health system. Skills in analysis are not only important for researching evidence of change as a result of interventions or treatments, they are also a vital component of a service committed to continuous improvement.

Evaluating program initiatives

Table 2.2 shows, for the hospital as a whole, the results on the key organizational health indicators for both years of the survey, together with the amount of change that occurred between the two administrations of the survey. A series of paired sample *t*-tests showed that there was a significant improvement in all measures except role clarity and individual distress. These two measures did not show a statistically significant change across the two years of the survey.

Table 2.2 Mean levels of each climate dimension across the hospital for both years of the survey

Survey dimension	Year 1	Year 2	Change
Leadership	51.6	53.9	2.3
Role clarity	65.6	65.8	0.2
Professional interaction	58.8	62.8	4.0
Professional development	47.0	51.6	4.6
Goal congruence	53.7	59.8	6.2
Recognition and appraisal	42.0	44.7	2.7
Participation in decision-making	42.6	48.6	5.9
Excessive workload	59.8	53.9	−6.0
Workplace morale	48.2	55.4	7.2
Workplace distress	56.7	51.3	−5.4
Individual morale	59.4	61.3	1.9
Individual distress	34.9	32.9	−2.0

Note: Scores can range from a minimum of 0 to a maximum of 100.

The staff evaluated the program initiatives that were developed after the second wave of the survey. Two types of evaluation were conducted. First, all staff in the workgroup rated the effectiveness of interventions. Second, a panel of staff representatives who were involved with strategic level interventions rated the effectiveness of interventions in each workgroup. Ratings were made on six aspects of the change initiative: magnitude, effectiveness, novelty, benefit to patients, benefit to staff, and benefit to administration. Each aspect of the initiative was rated on a five-point scale where zero indicated the initiative was very low on that aspect and five indicated that it was very high. Ratings across the six aspects were summed to create an overall index of the impact of the initiative in each workgroup.

The mean rating for staff was 3.6 (SD = 0.6) and for the panel was 2.7 (SD = 0.5). The correlation between the workgroup ratings and the panel ratings was 0.16. Table 2.3 shows the correlation between the ratings of program effectiveness, ratings from both staff and from experts, and the survey dimensions. The table shows the correlation with the survey dimensions for the prior year and also for the change over the two years.

The low correlation between the two types of rating, together with the different pattern of correlation in Table 2.3, suggests that the two types of rating were influenced by different factors. Workgroup ratings were more closely linked to the climate dimensions of leadership, role clarity, and professional development. Change in climate was a better predictor than simply using the climate measure for the current year. Panel ratings

Table 2.3   Correlation of ratings of initiatives by staff and expert groups with climate dimensions for the current year and for change over two years

	Workgroups' own ratings of programs		Experts' rating of programs	
	Current year	Change over two years	Current year	Change over two years
Leadership	0.09	0.35**	0.16	0.07
Role clarity	0.24*	0.34**	0.02	−0.05
Professional interaction	−0.13	0.01	0.23*	0.13
Professional development	0.23*	0.42**	0.06	0.21
Goal congruence	0.00	0.09	0.21	−0.09
Recognition and appraisal	0.21	0.19	0.30**	0.24*
Participation in decision-making	−0.22	−0.10	0.23*	0.20
Excessive workload	0.10	−0.27*	−0.04	−0.15
Workplace morale	0.05	0.22*	0.41**	0.26*
Workplace distress	0.02	−0.22*	−0.22*	−0.30**
Individual morale	0.48**	0.53**	0.01	0.15
Individual distress	−0.38**	−0.25*	−0.09	−0.21

* $p < 0.10$ ** $p < 0.05$

were more closely linked to the climate dimensions of appraisal and participation in decision-making and these correlations were weaker when change over time was the predictor.

Another key difference was that individual morale and distress were the most important predictors for workgroup ratings but workplace morale and distress were the most important predictors for panel ratings. This difference suggests that workgroups were influenced more by their own experiences while the panel ratings were influenced by factors more easily observed by outsiders to the workgroup.

MANAGING ORGANIZATIONAL HEALTH: INTEGRATING ORGANIZATIONAL DIAGNOSIS AND INTERVENTIONS

The case study presented above described a systemic approach to managing organizational health in a large urban hospital. A key element of the approach was the integration of survey-based diagnosis with strategic change activities to ensure that the survey was part of ongoing hospital management. The diagnosis of organizational health provided information about the current state of the organization in terms of core processes and employee well-being. The benchmarks and mathematical modeling provided a comprehensive 'snapshot' of organization health in the hospital. The 'snapshot' was quite detailed and, in the second year, included information about the way the workplace had changed. However, no matter how much detail is provided in a diagnosis, it remains a picture of what has happened up to the point of the diagnosis. To become part of ongoing change, information from the diagnosis must influence decision-making and action within the organization.

A key element of the success of the project in the hospital was that the survey was not treated as an optional or a separate activity. The survey was part of ongoing management throughout the organization. This integration was achieved by making sure that the survey data and responses were part of broader programs to improve organizational health. Integration was also achieved by linking the survey data to other information systems within the hospital. In summary, the survey was integrated with ongoing management by:

- using mathematical modeling and benchmarks to develop strategic initiatives to improve organizational health;
- using benchmarks within workgroups and strategic results to develop local initiatives for improvement;
- coordinating information about the content and success of both types of initiative; and

- linking data from the survey to other information systems in the hospital.

The first level at which survey results were integrated with ongoing management was at the strategic level. At this level, mathematical modeling results, as well as benchmark information, were particularly useful for developing an overall plan for improvement. The mathematical modeling results helped to identify key drivers of employee outcomes and helped to prioritize areas for change. Decision-making at this level was also important for allocating resources to workgroups to assist in the development of local initiatives.

For local initiatives, it was important that workgroups received specific feedback about the way climate factors worked within their workgroup. In this way, the workgroup was able to develop strategies that responded to the specific conditions upon which they could act more directly. At this level, benchmark information is particularly important because it provides a framework for identifying important issues and exploring their meaning in the specific workgroup. Workgroups were also provided with summary results from the mathematical modeling and the priorities that were established at the strategic level. The range of initiatives developed in the hospital included new communication strategies, participative decision-making forums, and restructuring teams. The evaluation of these local strategies indicated that workgroups could develop effective and meaningful interventions directed towards the needs of their particular workgroup.

It was also important to coordinate information about both the strategic level initiatives and the workgroup initiatives. Workgroups needed to be informed about the kind of organizational strategies and resources that were influencing change in the hospital. Similarly, senior managers needed to be aware of the kind of changes that were being implemented across the hospital. The coordination of this information was undertaken through both formal and informal channels. The development of specific implementation teams comprising representatives of professional and administrative groups was critical for coordinating this information.

The final aspect of successful integration of the survey is linking survey information to other information systems in the organization. The hospital implementation team is integrating climate measurement with human resource (HR) system data such as absence costs, employee health and safety, and customer service measurement. Climate results for workgroups can then be linked to relevant outcomes to identify issues of concern and to monitor the effectiveness and consequences of organizational change.

Each of the above factors required commitment from senior management to the overall process. We summarize the key elements of the overall process in Figure 2.4. The figure depicts the cyclical link between survey diagnosis and organizational intervention to improve organizational health.

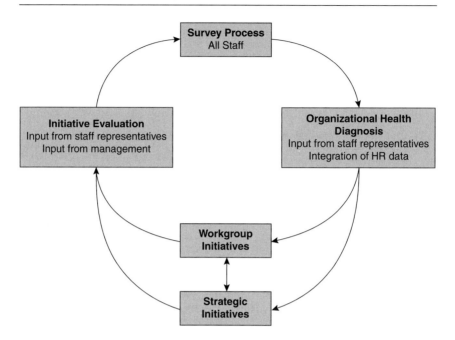

Figure 2.4   Cyclical process of diagnosis and intervention

Conclusion

Improving organizational health requires organizations to understand the relationship between employee well-being and organizational effectiveness. In a healthy organization these two outcomes are mutually supportive. Aligning individual and organizational outcomes to promote organizational health can then become a factor leading to long-term organizational success. Improving organizational health also requires understanding of the individual and organizational factors that contribute to valued outcomes in the organization.

Organizational climate is a key driver of employee well-being end organizational outcomes. Employee opinion surveys can provide a comprehensive diagnosis of organizational climate that can be integrated into the day-to-day management of the organization. This integration requires commitment from all staff and specific strategies to link survey diagnosis to the implementation and evaluation of interventions in the organization.

Note

1   This research was supported by the Australian Research Council (Grant No. A79601474) awarded to the first two authors and by a National Health and

Medical Research Council Public Health Fellowship (Grant No. 954208) awarded to the second author.

References

Barrick, M. R. and Mount, M. K. (1991) The big five personality dimensions and job performance: a meta-analysis, *Personnel Psychology* 44: 1–25.

Bolger, N. (1990) Coping as a personality process: a prospective study, *Journal of Personality and Social Psychology* 59: 525–37.

Costa, P. T. Jr and McCrae, R. R. (1980) Influence of extraversion and neuroticism on subjective well-being, *Journal of Personality and Social Psychology* 38: 668–78.

Cox, T. (1988) Organizational health, *Work and Stress* 2: 1–2.

Cuttance, P. and Ecob, R. (eds) (1987) *Structural Modeling by Example: Applications in Educational, Sociological and Behavioral Research*, New York: Cambridge.

Edwards, J. R. (1992) A cybernetic theory of stress, coping and well-being in organizations, *Academy of Management Review* 17: 238–74.

George, J. M. (1996) Trait and state affect, in K. R. Murphy (ed.) *Individual Differences and Behavior in Organizations* (pp. 145–71), San Francisco: Jossey-Bass.

Hart, P. M. (1994) Teacher quality of work life: integrating work experiences, psychological distress and morale, *Journal of Occupational and Organizational Psychology* 67: 109–32.

Hart, P. M. (in press) Predicting employee life satisfaction: a coherent model of personality, work and nonwork experiences, and domain satisfactions, *Journal of Applied Psychology*.

Hart, P. M., Griffin, M. A., Wearing, A. J. and Cooper, C. L. (1996a) *QPASS: Manual for the Queensland Public Agency Staff Survey*, Brisbane: Public Sector Management Commission.

Hart, P. M., Wearing, A. J. and Griffin, M. A. (1996b) *Integrating Personal and Organizational Factors into a Coherent Model of Occupational Well-being: A Covariance Structure Approach*. Paper presented at the Academy of Management Meeting, Cincinnati, Ohio.

Hart, P. M. and Wearing, A. J. (1999) Using employee opinion surveys to identify control mechanisms in organisations, in W. J. and A. Grob (eds). *Control of Human Behaviour, Mental Processes and Consciousness* Mahwah, NJ: Lawrence Erlbaum.

Hart, P. M., Wearing, A. J. and Conn, M. (1995a). Conventional wisdom is a poor predictor of the relationship between discipline policy, student misbehaviour and teacher stress. *British Journal of Educational Psychology* 65: 27–48.

Hart, P. M., Wearing, A. J. and Headey, B. (1995b) Police stress and well-being: integrating personality, coping and daily work experiences, *Journal of Occupational and Organizational Psychology* 68: 133–56.

Hart, P. M., Wearing, A. J., Conn, M., Carter, N. L. and Dingle R. K. (1999) *Development of the School Organisational Health Questionnaire: A Measure For Assessing Teacher Morale and School Organisational Climate*. Manuscript under review.

Hart, P. M., Wearing, A. J., Leipens, I. and Griffin, M. A. (1997) *Linking Organizational Climate to Employee Well-Being and Performance: A Covariance Structure Approach.* Paper presented at the 12th Annual Conference of the Society for Industrial and Organisational Psychology, St Louis, Missouri.

James, L. R. and McIntyre, M. D. (1996) Perceptions of organisational climate, in K. Murphy (ed.) *Individual Differences and Behavior in Organisations* (pp. 416–50), San Francisco, CA: Jossey-Bass.

Jimmieson, N. and Griffin, M. A. (1998) Linking staff and client perceptions of the organization: A field study of client satisfaction with healthcare services, *Journal of Occupational and Organizational Psychology* 71: 81–96.

Kraut, A. I. (ed.) (1996) *Organizational Surveys: Tools for Assessment and Change,* San Francisco, CA: Jossey-Bass.

Lazarus, R. S. (1990) Theory-based stress measurement, *Psychological Inquiry, An International Journal of Peer Commentary and Review* 1: 3–13.

Lazarus, R. S. and Folkman, S. (1984) *Stress, Appraisal, and Coping,* New York: Springer.

Michela, J. L., Lukaszewski, M. P. and Allegrante, J. P. (1995) Organizational climate and work stress: a general framework applied to inner-city schoolteachers, in S. L. Sauter and L. R. Murphy (eds) *Organizational Risk Factors for Job Stress* (pp. 61–80). Washington, DC: American Psychological Association.

Miller, R., Griffin, M. A. and Hart, P. M. (1999) Personality and organizational health: the role of conscientiousness, *Work and Stress* 13: 7–19.

Neal, A. and Griffin, M. A. (1998) *Safety Climate and Safety Related Behavior: An Integrated Model.* Paper presented at the 24th International Congress of Applied Psychology, San Francisco, California.

Schneider, B. (1990) The climate for service: An application of the climate construct, in B. Schneider (ed.) *Organizational Climate and Culture* (pp. 383–412). San Francisco, CA: Jossey-Bass.

Chapter 3

Improving organizational effectiveness through integration of core values

D. Fassel, J. Monroy, and M. Monroy

INTRODUCTION

This chapter describes an approach to organizational change that focuses on enhanced commitment to core values as the means to improve organizational effectiveness and worker well-being. The approach involves an assessment of organizational functioning, measurement of how well the core values are being demonstrated in everyday work life, and identification of key drivers for improving commitment to core values, worker well-being, and organizational effectiveness.

The organization under discussion is a not-for-profit healthcare system that employs 45,000 people in 50 facilities throughout seven geographic regions, most of which are located in a single western state. The organization is growing rapidly, expanding its market presence through the acquisition of small and/or independent healthcare. This expansion is in keeping with the organization's mission of providing quality, affordable healthcare, especially to the poor. As is the case with all healthcare facilities today, the facilities within this organization are increasingly pressed by consumer demands for quality, low-cost healthcare, increased regulation, and shrinking insurance reimbursements.

In the spring of 1998, the corporate leadership of this healthcare system determined that the best strategy for achieving the organization's mission was to fully integrate five core values throughout all work sites. These core values were not new to the organization. However, this was the first attempt to measure the degree to which they are integrated, as well as their influence on organizational effectiveness. In light of the expanding number of member hospitals, it was further believed that integrating the core values would provide a means of creating a common culture throughout the system. The core values are as follows:

- *Dignity* (respecting the inherent value and worth each person possesses as a member of the human family).

- *Collaboration* (working together with people who support common values and vision to achieve goals).
- *Justice* (advocating for change of social structures that undermine human dignity, demonstrating special concern for those who are poor).
- *Stewardship* (cultivating the resources entrusted to our care as we promote healing and wholeness).
- *Excellence* (a shared commitment to exceed expected results in our work and services through teamwork and innovation).

Core values and attention to ethical development are not new in organizational life. Indeed, there has always been the tacit assumption that organizations are large "moral persons" or entities that are bound by the laws of the land and by principles that guide their behavior with customers and with employees. More recently, as society has become more fragmented and pulled in many directions, there has been a renewed emphasis on values in many areas of life. Family values have become a political rallying cry; programs that teach values in schools have been funded and implemented; and now there is a heightened interest in values in organizations.

All organizations have core values, whether stated or unstated, and they function like the soul of the organization. Values are an internal element that provides guidance and inspiration. In themselves, values are invisible, and only their demonstration via action and their effects can be observed. When values truly live in an organization, they can contribute mightily to organizational effectiveness. For example, in their study of successful companies, Collins and Porras (1994) found that visionary companies are driven by a core ideology that consists of core values and a purpose. Further, these companies were characterized as premier institutions in their industry, widely admired by knowledgeable business people, having made an indelible imprint on the world; they were still vital 50 years after their founding. Their definition of core values is similar to other definitions found in the literature and suits the purposes of this article. They say that core values are ". . . the organization's essential and enduring tenets – a small set of general guiding principles; not to be confused with specific cultural or operating practices; not to be compromised for financial gain or short-term expediency" (Collins and Porras: 73).

Core values are typically short, probably no more than five or six in number, and they do not come and go with the trends and fads of the day. They are authentically "core" in that they would not change even if the business environment penalized a company for holding such values. Corporate executives who became aware of the work of Collins and Porras in this area were drawn by something else in addition to the obvious rightness of having values. They were drawn by the fact that their study of successful companies showed that core values were good for business. They demonstrated, through detailed examination of financial records, that visionary companies – those with a purpose and core values – were much more financially successful over the course of their existence. Obviously, core values were good for the bottom line.

Collins and Porras weren't the only ones giving voice to the importance of values. In a remarkable case study of the turnaround of Sears led by CEO Arthur Martinez, Sears undertook a complete revamping of the way it did business. Sears involved 80,000 employees in identifying six core values that led to improved employee satisfaction, which, in turn, led to increased customer satisfaction and, finally, to greater profit (Rucci *et al.* 1998). Clearly, the movement in the values arena was not just one of articulating values, but of integrating values so that they permeated every aspect of work, all the way up and down the chain. Additional examples of companies who try to "live out" their core values can be found in Anderson (1997).

If commitment to core values influences organizational effectiveness, then it should be possible to demonstrate this relationship by measuring the effect of values and values integration on organizational climate and organizational effectiveness. That is, to determine whether values change the way people feel about their organization, do their work, make decisions, and interact with customers.

It was the confluence of these issues, the importance of values and the integration of values into the workplace, the impact of values on organizational effectiveness, and the necessity to measure that impact, that led to the current study.

METHODS

The traditional approach to employee satisfaction involves distributing a lengthy questionnaire to employees every two or three years. The survey is often called an "employee satisfaction" or "climate" survey. The purpose of the survey is to collect information on a wide variety of work-related topics and to report the results by department/unit level, as well as by gender, years of service, race, etc. The data are analyzed and eventually fed back to the organization in a series of charts and tables in which the usual analytic unit is a Percent Favorable score. Each demographic category, as well as each department/unit, receives a list of the top 10 and the lowest 10 Percent Favorable scores. The questions with the lowest scores are targeted for change.

Often, employees never see the data; instead, managers use the data to chart new change initiatives for their department/unit and pass down these initiatives to employees in the form of new policies or "activities" designed to improve efficiency. It is not unusual for these change efforts to take many months to develop. Thus, the link to the questionnaire assessment becomes more and more tenuous as time passes. Change efforts usually "lose steam" over time and, after a few months, they disappear completely. A year or so after the change initiative has stalled the consultants reappear and announce that it is time to take the survey again.

A new approach

The healthcare system under discussion had conducted traditional employee surveys once every three years. Now the organization's leadership sought a survey that would systematically measure the integration of the core values at each facility, as well as provide a clear focus on the specific organizational behaviors with the strongest statistical linkages to the integration of the values. It was hypothesized that committing resources and working to improve these identifiable organizational behaviors would improve the extent to which each of the core values is integrated, and ultimately would improve overall organizational effectiveness.

The usual approach to conducting, analyzing, and feeding back employee survey results was abandoned in favor of a different strategy. Newmeasures, Inc. developed a single-page 16-item survey called Developing Organizational Capacity (DOC©) which was designed for distribution to employees every six months. In this instance, the basic 16-item survey was expanded to included five additional items that measured the degree to which the organization's five core values were integrated. Managers received a full report within two weeks. They were shown specifically where to focus their change efforts to improve the practice of the core values, and ultimately to improve organizational effectiveness. A shorter employee report was also prepared. A single focus for change was provided for the entire facility; no breakdown reports were prepared for individual departments or units within the facility. Thus, all employees at the site focused on a single change initiative.

Development of the survey

The DOC© survey was developed over a two-year period, between 1994 and 1996. The goal was to develop a brief yet reliable measure of organizational functioning, a one-page survey that could be answered in less than five minutes. The rationale was as follows: today's organizations do not have the luxury of allocating large amounts of employee time to take opinion surveys, wait months to have the data fed-back, and even longer before ideas for improvement are discussed and implemented. Organizations need a "real time" snapshot of their operating systems and specific, immediate ideas for improving effectiveness and the quality of work life.

We began with a 120-item survey from a Fortune 200 company and used factor analysis and multiple regression to select subsets of reliable and valid questions. In a series of analyses, we sequentially reduced the number of questions until we arrived at the "best of the best." The final set of 16 questions measured a critical set of organizational factors, which have been identified as important in hundreds of research studies. These factors include communication, decision-making, innovation and communication, among others. The survey also measured factors at all organizational levels: the worker, job/task, workgroup, supervisory, and culture/climate.

A note on single-item scales

The use of single items to measure key organizational constructs requires some explanation. Most organizational factors are complex and have multiple facets or dimensions and researchers traditionally have used multi-item scales to measure these types of factors. However, the problem with single-item scales is lessened considerably when one has experience with particular items and can compare the precision of single-item measures with multi-item measures. In the present case, we could compare how much of the total scale variance a single item explained (i.e. how well it measures the construct) and how the single-item and multi-item scales correlated with other constructs of interest (e.g. turnover intent, job satisfaction, product quality, etc.). In our analyses, the multi-item scales routinely had higher correlation with various constructs of interest than the single-item measures. This was not surprising. What was noteworthy was that the pattern of differences among the two sets of correlations was not high, and an omnibus F-test of all the set of differences was not statistically significant. Thus, the single-item scales were found to be conservative estimates of the true correlation (as judged by the multi-item scale). So, what was lost in the trade-off with size is a small degree of precision.

Survey procedures

Key contact persons within each region and at each facility were identified, usually human resource directors and managers. Their role was to assist in the collection of data and the reporting of results. They were given background information on the survey's purpose and procedures, which emphasized the shortness and ease of using the survey (three to five minutes to complete) and the fact that it was not a traditional employee satisfaction or climate survey. Rather, the survey was an instrument to measure how well the organization's core values were integrated into the workplace. The survey process was described in non-threatening terms as a new approach to organizational health that would be repeated every six months, with the goal of its becoming a routine part of the organization's reporting, much as financial reporting is done on a routine basis (monthly or quarterly).

Surveys were distributed in various ways: as paycheck stuffers, by supervisors within workgroups, and at shift/communication meetings. The logistics of returning the surveys were also managed in various ways at each facility. These survey returns ranged from boxes placed in employee break areas and cafeterias to self-addressed, stamped envelopes, which employees could mail directly to the independent company contracted to enter the survey data.

At the majority of the facilities where employees did not mail their survey responses directly, key contact persons were responsible for keeping the process moving forward and seeing that the completed surveys were bundled by facility and mailed on schedule for data processing. Typically, the surveying period lasted one week. The average response rate to the survey was 62 percent; among full-time employees, the rate of participation was nearly 75 percent.

Feedback of results

Within two weeks of the survey's completion, the results were sent to the key contact person at each facility. In most instances, an initial review of the results was carried out in a phone conversation with this person. With the contact person, we established a date for onsite feedback, which consisted of a presentation to the facility's leadership, as well as meetings with any/all of the following: task force groups formed to implement changes, groups of supervisors, and employees.

Results were communicated openly among employees at each facility. Teams were formed to develop action plans to address the one or two key organizational behaviors identified by the survey data as having the strongest statistical linkage(s) with the core values and overall organizational effectiveness. This process would be repeated every six months.

In addition, we met twice a year with key contacts from each region to review the process, identify any overall trends, report on change efforts, and encourage benchmarking. From these meetings, a benchmarking memo was developed and distributed to encourage facilities to share best practices and, whenever feasible, to take a systems approach to change efforts. The healthcare organization system office also dedicated one organizational development person to act as a resource to help facilities design their change efforts.

RESULTS

More than 20,000 workers from 50 different hospitals and related healthcare facilities completed the DOC© surveys. Response rates to the survey ranged from 45 percent to 75 percent and the average response rate was

62 percent. The results are presented in two complementary parts. Part 1 describes the analyses that we performed for each facility, using the results for one facility as an example. Part 2 presents a multivariate analysis (canonical correlation) of all 20 facilities combined to identify those organizational factors associated with both employee well-being and organizational effectiveness across the facilities.

Part I: Individual facility analyses

For each facility, the data were analyzed in five steps. First, the percentage of workers who gave favorable (agree or strongly agree), neutral (neither agree nor disagree), and unfavorable (disagree or strongly disagree) scores were calculated for each question. This represents the traditional manner in which survey data are analyzed and results fed back to management and workers. While Percent Favorable scores provide information on whether the score for each item is high or low, they do not show how the item is related to organizational effectiveness. A more sophisticated statistical tool, such as correlation analysis, is needed to determine the relationship between each item, organizational effectiveness and the core values. If the relationship is small, then it would not be helpful to focus change efforts on that area, even though the Percent Favorable score is low. In our analyses, we provided Percent Favorable scores but did not rely on them alone for the organizational diagnosis.

Next, a work systems analysis was performed. Scores on selected items were used to represent six major work systems: Rewards; People; Tasks; Information; Decision-making; and Structure. One set of Percent Favorable scores for each of the six work systems was calculated for the entire sample, and a second set of scores for those who reported coping well with work pressure. Both sets of scores were plotted on a single chart as shown in Figure 3.1.

Differences between the two lines were examined with special reference to those that exceeded 20 percent. Any such differences were thought to reflect problems within that work system and an indicator that corrective action should be taken. (A useful analogy is that of the canary in the mine: miners used canaries to provide early warning of the presence of noxious gas.) By and large, such differences between groups of good copers and the entire facility were uncommon. What were common were very low scores on the reward question, both for the entire sample and for those who were coping well. The extent to which these low scores reflected financial issues or lack of recognition is unclear, and additional work is currently under way to disentangle these two possible explanations.

The third analysis assessed the reliability of the five "value" items using coefficient alpha. This analysis indicates the extent to which the five value

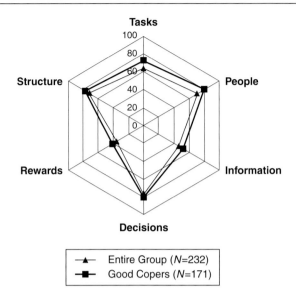

Figure 3.1   Work systems analysis for one healthcare facility

items form a uni-dimensional scale (i.e. whether they are measuring the same construct). In each facility, the value items formed a single, uni-dimensional scale with coefficient alphas ranging from 0.88 to 0.93. This indicates that workers respond to each of these items as if they measured a single construct.

The fourth step in the analysis was to measure the statistical relationship between the five core values and organizational effectiveness. The range of coefficients was from 0.58 to 0.70, with an average of 0.60. Our findings were consistent with prior research on the important relationship between an organization's values and its effectiveness.

The fifth and final step in the analysis was to identify the key "driver" for improving organizational effectiveness via better value integration. The intent was to provide the facility with the one or two best areas on which to focus their change efforts. The facilities were not interested in detailed structural models showing many factors that could be changed. Rather, they were interested in one or two areas to focus their attention and energy, which would improve value integration. Accordingly, the value integration scale was "forced" into the statistical model in the position adjacent to organizational effectiveness, and correlation analysis was used to find which survey items correlated most highly with the value integration scale. Once the single or two best drivers were found, a path model was constructed using multiple regression.

Figure 3.2 shows the path model for one facility. The model reveals that, in this case, the pathway to improved organizational effectiveness begins with the following "driver": coached to improve performance. Reading the path model from left to right this translates to: improvement in "coaching to improve performance" should lead to improvements in workers feeling valued and in value integration, which in turn should foster higher organizational effectiveness.

It should be noted that in most of the facilities tested, the final path model explained a minimum of 50 percent of the variance in organizational effectiveness. This indicates that knowledge of how workers score on the drivers allows us to predict value integration overall effectiveness with a high degree of confidence.

While the patterns of results varied across the facilities, some common areas for improvement emerged. For example, the most common driver identified across the facilities was "coached to improve performance," which occurred in more than 60 percent of the facilities. Because of its prominence in many of the path models, efforts to improve coaching are under way at these facilities. In developing tactics to improve coaching, we noted that in the research and consulting literature, coaching was typically associated with organizational leadership or "high potentials." By making coaching available to everyone in the system, the facilities are taking an innovative step toward improving value integration and organizational effectiveness. This innovation offers promise for producing large, system-wide improvements in the second round of surveys.

Two other drivers were identified in about 25 percent of the facilities' path models. One was "encouraged new ideas." This driver was strongly linked statistically with another survey question: "workers have decision-making authority." These two items seem to be measuring the climate for innovation and creativity. The other driver was "rewarded for performance." This is somewhat difficult to interpret because it contains elements of formal pay, but also of recognition and/or praise. Additional research is being conducted to disentangle these elements.

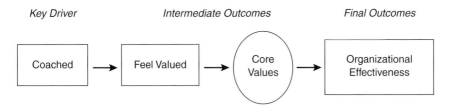

Figure 3.2 Sample model from one healthcare facility showing the pathway for change to improve organizational effectiveness and employee well-being

Part 2: Multivariate analyses

A second set of analyses was performed to examine overall patterns in the data, with special reference to the links between effectiveness and employee well-being. In the context of a healthy work organization model, the analyses would show which characteristics of the organization were correlated with both effectiveness and employee well-being. Once identified, these characteristics become the targets for change efforts to create a healthy work organization. For these analyses, all 50 facilities were combined and a canonical correlation analysis was performed. Canonical correlation examines relationships among multiple independent and dependent variables. There were four dependent variables: organizational effectiveness, workers feeling valued, work pressure, and coping with work pressure. The first variable was a performance indicator and the other three were measures of worker well-being. The independent variables were the commitment to values scale, plus the remaining 13 survey items which dealt with a variety of factors including communication, innovation, workload, cooperation, conflict resolution, supervisory support, and safety and health conditions. These analyses would indicate the extent to which the dependent variables could be predicted by the independent variables.

The results indicated that the first canonical variate was very large with an adjusted canonical correlation of 0.81; the remaining canonical varieties were much smaller, having correlations of less than 0.30. The first canonical variate accounted for 42 percent of the variance in the four outcome variables. Inspection of the canonical structure of the independent variables revealed that the strongest contributors were (in order of size): commitment to values, employee coaching, encourage new ideas, and rewards for performance. These four factors were the most strongly related to both worker well-being and organizational effectiveness outcomes. These are the same four factors which appeared in many of the univariate analyses performed for the individual facilities.

Finally, it should be noted that the four dependent variables were not equally well predicted by the independent variables. For example, about 56 percent of the variance in organizational effectiveness and feeling valued could be explained, while about 26 percent of the variance in work pressure and good coping could be explained. This suggests a need to identify worker well-being variables other than stress and coping for inclusion in future healthy work organization studies, which would allow the predictor variables to explain a larger share of the variance in outcome measures.

Figure 3.3 provides an illustration of how the key factors from the multivariate analyses link together in a path diagram. This model allows the organization to do two things. First, it provides guidance on how to design targeted interventions to improve worker well-being and

Key Driver   *Core Values*   *Final Outcomes*

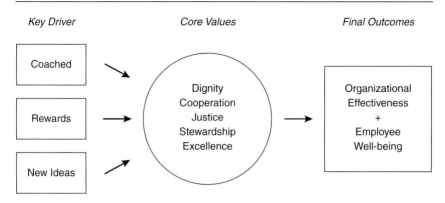

Figure 3.3 Summary model showing pathway to improvement of organizational effectiveness and worker well-being (20 healthcare facilities)

organizational effectiveness. Second, it directs which types of variables need to be measured to assess the efficacy of planned interventions.

Link to performance and financial outcomes

Future work will involve linking the employee survey data to organizational performance and financial outcomes. We developed a model showing how the process of value integration would improve customer satisfaction and financial performance in healthcare organizations. The model specifies that the main effects of the value integration process will be on changes in organizational practices, such as improved communication, empowering workers to improve the quality of work, and fostering supervisory support. The changes in organizational practices will drive employee perceptions of management commitment to the core values, which will have two effects. The primary effect will be on employee outcomes (such as feeling valued, job satisfaction, teamwork), but a secondary effect on service outcomes also is predicted. However, the main effects of commitment to values on service outcomes will be mediated by the improved employee outcomes. The improved employee and service outcomes will lead, respectively, to organizational effectiveness and patient satisfaction which in turn will influence financial results (market share, return on equity).

Employee outcomes are central to the model and moderate the effects of organizational practices on both customer satisfaction and financial performance. If tests of the new model prove positive, then the case for organizations to invest time and energy to actively create healthy and productive work environments becomes quite strong and there would be little reason not to initiate such efforts.

DISCUSSION

The results of this study revealed a clear linkage between leadership commitment to core values and organizational effectiveness. It is important to note that having core values and demonstrating commitment to them are two distinct issues. Only the latter is associated with improved effectiveness and employee well-being. Many organizations advertise their values and hold them up to employees and the local community. This is the easy part. The hard part is to align decision-making at all levels with the core values. This includes decisions about hiring, firing, layoffs, training, employee development, purchasing, and maintenance. At each point, alignment with the core values should be checked. Anderson (1997) offers some thoughtful comments about decisions that involve conflicts among and between core values and financial (bottom-line) issues.

The importance of leadership commitment to stated beliefs and policies should not be surprising; it has been highlighted in other areas of research. For example, it is well known from the occupational safety literature that leadership's commitment to safety is a key element in successful safety programs (Cohen, 1977). In this classic study of high- and low-accident companies, low-accident companies were characterized as providing performance feedback to employees, having a humanistic management style, high worker involvement in safety, good hygiene practices, and stable employment practices. Building safety into everyday work activities has become a tried and true way of reducing accidents and fostering safe work behavior.

More recent studies have confirmed these results. Leadership commitment to safety was a key driving force in a study of compliance with universal precautions among healthcare workers (Gershon et al. 1995). Job-related factors such as workload and stress, and personal factors like risk-taking attitudes were far less predictive of compliance with recommended work practices than were leadership commitment to safety and performance feedback.

Are the hospital facilities described in this chapter healthy work organizations? The answer is not clear yet, but it is certain that the facilities are moving in the right direction. Efforts are under way to improve commitment to the core values and to integrate these values into the everyday work experience. Surveys to be taken every six months over the next few years will provide indications of how well the changes are working and what additional changes may be needed to improve organizational effectiveness and employee well-being. Repeating the surveys every six months sends the message to managers and employees that the change efforts are not short term and will not "die on the vine." Employees will see that the organization is serious about the desire to change and to live out their core values at work. This level of commitment on the part of the organization, if sustained, should eventually result in positive changes in those organizational practices which foster both improved worker health and organizational effectiveness.

References

Anderson, C. (1997) Values-based management, *Academy of Management Executive* 11: 25–46.

Cohen A. (1977) Factors in successful occupational safety programs, *Journal of Safety Research* 9: 168–78.

Collins, J. and Porras, J. (1994) *Built To Last: Successful Habits of Visionary Companies*, New York: HarperBusiness.

Gershon, R. M., Murphy, L. R., Felknor, S., Vesley, D. and DeJoy, D. (1995) Compliance with universal precautions among healthcare workers, *American Journal of Infection Control* 23: 225–36.

Rucci, A., Kirn, S. and Quinn, R. (1998) The employee–customer–profit chain at Sears, *Harvard Business Review* (Jan–Feb): 82–97.

Chapter 4

Managing change in healthcare: a study of one hospital system's efforts

S. K. Moran and T. R. Parry

INTRODUCTION

The healthcare industry is experiencing tremendous change, and innovations continue in healthcare technology. Changes are occurring in reimbursement methods, patient record keeping, service re-engineering, and even patient expectations. There is a growing influence of consumer groups, and the medical delivery system itself is becoming more complex. Creating even more challenge and change are the downsizings, mergers, and acquisitions of healthcare organizations themselves (Sers 1998).

As if this turbulence was not enough, the criticality of accurate performance in medicine increases the burden placed on an employee for quality performance. Healthcare is an industry where tolerance for error is minimal (Bogner 1994). Performance decrements can result in declines in patient outcomes varying from a vegetative state even to death. The consequences of mismanaging change, therefore, are far-reaching and severe.

St Paul Medical Services and Health Management Associates (HMA)

A unique opportunity to study employee perceptions of managing change and important organizational outcomes within a healthcare system arose between St Paul Medical Services (St Paul) and Health Management Associates, Inc. St Paul Medical Services, part of the St Paul Companies Inc., is the United States' premier medical liability insurer. St Paul has provided many years of coverage for HMA, a leading operator of acute care hospitals in non-urban areas of the south-east and south-west United States.

HMA is an organization whose mission encompasses change. In the early 1980s HMA refocused its corporate mission to acquire underperforming non-urban hospitals and to provide the highest quality medical care and medical services in these rural markets. Their philosophy is based in part on the fact that limited access to basic or specialized medical services forces individuals to seek medical care far from home. For example, primary care physicians are often compelled to refer their patients to specialists in other towns. The local hospital loses admissions and outpatient procedures to more metropolitan facilities. The goal is to reattract patients to these rural HMA-acquired facilities. HMA directs a successful acquisition strategy in response to these competitive factors and individuals' desire to access healthcare in their own community.

HMA's acquisition strategy

Acquisition candidates for HMA must meet specific, demanding acquisition criteria: location in a high-growth, non-urban market, proven demographic need, community support for expanded services, potential to become the sole or dominant provider, underutilization by local physicians, strong primary care physician base, and an attractive purchase price.

Prior to closure, HMA seeks to gain the trust of the acquired facility. Hospital trustees and business leaders are encouraged to contact their HMA peers to learn more about HMA and to convince them that HMA is a financially successful employer and an excellent corporate citizen in the communities it serves.

Immediately following an acquisition, HMA applies a consistent management plan to maximize efficiency and improve operating margins. They introduce decentralized management with highly centralized operational monitoring, implement a proprietary, management information system to monitor financial and patient information, and evaluate and appoint without undue delay an experienced healthcare management team at the hospital level. They maintain strict supply cost control measures through volume-discount purchase agreements, empower the facility by allowing local decision authority and employee involvement, and improve and expand tertiary services to decrease outmigration and increase utilization.

To alleviate fears and concerns about workforce reductions and layoffs, HMA often commits in writing to no workforce reductions for one year post-acquisition (excluded from this agreement are "for cause" personnel terminations). Flexible staffing programs are introduced, and the transfer to HMA benefits, payroll, and policies is made as transparent to employees as possible. An HMA human resources director is selected as a support person for the acquired facility. He or she works on-site at the new facility for a few days, reviewing policies and manuals, and answering questions at orientation meetings.

HMA's access to capital and commitment to aggressive physician recruitment helps to reduce outmigration. For the first five years post-acquisition, HMA infuses about two million dollars annually into the facility (Blecher 1998). State-of-the-art equipment is installed, services are expanded, and specialty services are added. Physicians are actively recruited and offered reasonable financial assistance to relocate to the community. By redefining the local specialty mix of physicians and services, HMA strives to provide a comprehensive range of medical services and directs the utilization of those services back into the community.

Today the company operates 32 facilities in 11 states with 3821 licensed beds. HMA continues to meet or exceed financial performance goals and is recognized as having the most efficient operating margins and the strongest balance sheet in the industry. Over the past five years, HMA has generated compounded annual EBITDA growth of 31 percent, among the best in the industry (Health Management Associates, Inc. 1998a). For the fiscal year ended September 30, 1998, net earnings expanded by 26 percent to $136.8 million compared to $108.3 million in fiscal 1997. HMA reported total net patient service revenue for the period of $1.14 billion, an increase of 27 percent from the prior year's figure of $895.5 million. The company has demonstrated 10 years of consistent and solid earnings growth (Health Management Associates, Inc. 1998b).

Through successful acquisitions of rural facilities, HMA seeks to provide high-quality healthcare services with a substantial financial growth. Given the amount of change within the healthcare industry itself, as well as within an organization that seeks such substantial growth, HMA was a prime candidate for the current study. And, because St Paul has a vested interest in the health, well-being, and financial success of its insured customers, it partnered with HMA to conduct a study of the impact of its acquisition efforts and the attempt to manage the resulting change.

Organizational outcomes

Organizational change and chaos have been studied in relation to a multitude of variables. For the healthcare industry, one common measure of organizational or facility success is quality. In addition to cost-effectiveness indicators, patient satisfaction data are now being used to make decisions such as choosing providers, determining bonuses, contract negotiations, even making changes in hospital food services. Bond (1997), for example, reported on efforts to improve satisfaction with one hospital's food services while at the same time reducing the average cost of patient trays. In a study of managed care organizations, Zimmerman and Zimmerman (1997) found that nearly half of the interviewed organizations said they would drop providers who failed to meet customer service standards. Other organizations, such as the Pacific Business Group on Health, are publishing patient satisfaction ratings of medical groups and physician networks (Larkin 1997).

The rationale for such studies is that satisfied and dissatisfied patients will behave differently. In fact, the link between patient satisfaction and important outcomes is similar to the one established in other industries with customer satisfaction (see Wiley 1996, and Schneider 1991). With respect to patient satisfaction, patients who are satisfied with their doctor and treatment, for example, are more compliant, and more likely to remain with their doctor, keep their appointments, and use services (Finkel 1997).

Another important organizational outcome factor is employee turnover. Dysfunctional turnover, that is, losing high-quality employees you would rather retain, is costly for an organization. The determinants of turnover are multifold; for example, involuntary turnover due to layoffs or job performance factors, and voluntary turnover from job dissatisfaction, current economic conditions, and the likelihood of finding another job (Mowday *et al.* 1982). Whatever the reason, and whatever the type, the consequence of high levels of turnover are troublesome for an organization. In healthcare, where so much of the care delivery hinges on team coordination and communication, staffing upheavals can have a dramatic impact (Helmreich and Schaefer 1994).

The present case study investigated organizational change efforts in 26 HMA facilities and sought to determine how these efforts relate to the two important outcomes of patient satisfaction and employee turnover.

METHODS

Sample

Hospitals owned by Health Management Associates (HMA) during 1995 and 1997 participated in the study. Participation was restricted to hospitals owned for at least six months prior to the time of survey in 1995 and 1997. As part of their continuing quality improvement efforts, HMA hospitals participated in St Paul's Human Factors Inventory. This inventory is described in greater detail below. Participation of individual employees in the survey was voluntary.

Measure of change management

The Managing Change scale from St Paul Fire and Marine's Human Factors Inventory was used to assess employee perceptions of the facility's efforts to manage change. The survey was administered to full-time employees of HMA facilities in the spring of 1995 and again in the spring of 1997 (see Figure 4.1). Employees responded anonymously to the survey by placing their answers on computer scan sheets. The six survey items for the Managing Change scale are: 1) this organization adequately informs employees about changes that will affect them; 2) this organization clearly communicates the reason behind changes that affect you; 3) this organization does a good job of helping employees adjust to changes at work; 4)

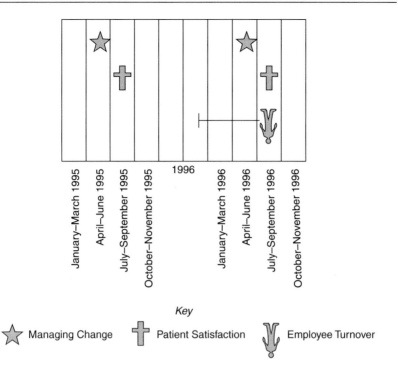

Figure 4.1 Timeline for variable measurement

this healthcare organization adjusts well to change (in the market, technology, etc.); 5) this organization does a good job preparing for the future; and 6) do you fear you will lose your job due to future changes within the organization? (reverse coded). Items were rated on a five-point Likert-type scale ranging from strongly agree or always, to strongly disagree or never. Higher scores indicate poorer change management efforts. The Managing Change scale has a demonstrated coefficient alpha reliability of 0.80 (St Paul Fire and Marine Insurance Company 1995).

Patient satisfaction

HMA participates in patient satisfaction surveys as part of their ongoing efforts for total quality management. The method of collecting patient satisfaction surveys was to distribute the surveys to patients just prior to discharge, along with any other papers relevant to the discharge procedure at each hospital. The discharge nurse was available to answer any questions the patient had about the survey. Patients were instructed to seal their completed survey and drop it in a prominently displayed "ballot type" box. Providing a name and address was optional.

Once a week, all surveys were reviewed by the Quality Services Management Director. Patient comments and recommendations were noted. The forms were batched and forwarded to a research company where they were scanned and verified, and the data were entered into a cumulative database.

All hospitals were to collect a minimum of 300 completed questionnaires per quarter, or an amount equivalent to 30 percent of inpatient discharges and outpatient surgical procedures, whichever was lower. A single composite percent figure was used to measure the overall quality of care and service delivery. The overall rating was computed by averaging the percent figures for each of the contributing questions, in proportion to the number of patients answering. The questions that comprise this composite are listed in Table 4.1. Fourth quarter ratings were available for both 1995 and 1997, so these were used to represent Patient Satisfaction for analysis.

Turnover

HMA facilities record employee turnover using the Bureau of National Affairs formula: the total number of employee separations during the year divided by the average number of employees on the payroll for that year. The percent of labor turnover is obtained simply by multiplying by 100. HMA defines turnover as any and all separations, whether voluntary or involuntary. The average number of employees on the payroll is determined by combining the total number of active and inactive employees divided by the number of pay periods in the year. For purposes of this study, turnover data for the fiscal year 1997 (October 1, 1996 through September 30, 1997) was used. Turnover data for 1995 had not been submitted to corporate, so they were not available for analysis.

Employee involvement and managing change strategies

HMA has an elaborate human resource plan for acquired facilities. Immediate post-acquisition activities include intensive meetings between corporate human resource directors and facility human resource personnel. Control of human resource systems, however, remains with the facility.

In an attempt to manage the human side of change and to promote effective human resource management, HMA requires all facilities to participate in a communication program they have called "Speakeasy." The goal of this program is to ensure that the hospital administrator and groups of employees periodically have a chance for direct communication. The hospital administrator meets with either a random group of employees or a selection of employees from various departments to discuss what is

Table 4.1   Questions comprising the overall patient satisfaction rating score by category headings

Overall Satisfaction
    In general, how would you rate the care you received?
Admissions/Business Office (Rate the service you received upon your admission)
    Promptness
    Courteousness
    Explanation of insurance/admission forms
    Explanation of billing procedures
General Nursing (Rate the care you received from the nursing staff)
    Did you receive assistance finding your way around the unit?
    Nurses showed concern and compassion
    Nurses personalized my care to meet my needs
    Questions were answered satisfactorily
    Did you receive discharge and follow-up care instructions?
    Did nurses identify themselves as your caregivers?
    Was your plan of care discussed with you?
Physician Care (Rate the care you received from your physician or physicians)
    Courteousness
    Explanation of pre-operative procedure
    Explanation of diagnosis
    Did you understand your discharge instructions?
Food Service
    Promptness of delivery
    Appearance of food and tray
    Overall taste
    Were the correct items delivered?
    Was hot food hot?
    Was cold food cold?
    If you were on a special diet ordered by your physician, did you receive an explanation?
Hospital Accommodations/Personal Comfort
    Courtesy of housekeeping personnel
    Were your room and bathroom cleaned regularly?
    Was your room temperature comfortable?
    Hospital's general cleanliness
Hospital Services (Rate the services provided to your during your stay)
    Intensive Care Unit
    Critical Care Unit
    Cardiac Rehab
    EKG
    Surgery
    Laboratory
    Obstetrics
    Physical Therapy
    Volunteers
    Respiratory Therapy
    Social Services/Discharge Planning
    Telephone Operator
    X-Ray
Operating Room Nursing Staff (If surgery was provided, rate the care you received from the operating room staff)
    Promptness
    Courteousness to you and your family
    Were your questions answered to your satisfaction?
    Did you understand your discharge instructions?

happening with HMA and with their facility. The hospital administrator answers questions employees may have, obtains suggestions for changes, and gathers employees' ideas. Each quarter, hospitals are required to summarize their Speakeasy activities in a report to corporate. The timeline for the variables measured in this study are summarized in Figure 4.1.

RESULTS

Twenty-one HMA facilities participated in the Human Factors Inventory (HFI) in 1995. The average participation rate of these facilities was 61 percent with 3900 usable answer sheets. An additional five facilities participated in 1997. The 1997 average participation rate was 90 percent with 7307 usable answer sheets (see Table 4.2). Answer sheets were usable if they met two criteria: the response sheet must contain 10 or fewer missing responses, and it must have an acceptable "distortion score." This separate HFI scale detects response patterns so unusual or overly favorable that they are unlikely to be valid. The scale is used to eliminate respondents who score above a preset cut-off. In general, this scale eliminates 0–5 percent of respondents and results in more accurate information for the organization. Because of this pre-screening process, the final survey results do not include responses from individuals who did not understand the survey or did not answer realistically. The percentage of HMA respondents eliminated because of a high distortion scale was 2 percent for both the 1995 and 1997 surveys.

Table 4.3 summarizes the means and standard deviations for managing change, patient satisfaction, and employee turnover.

Managing change

Recall that higher scores on the Managing Change scale indicate poorer change management efforts. In 1995, Managing Change scores for the 21 facilities ranged from 42 to 80 with a mean of 53.7. In 1997, for the 26 facilities, the scores ranged from 32 to 71 with a mean of 50.8. Although the 1997 Managing Change score is lower than 1995 (more favorable perceptions of managing change in 1997), it was not significantly so in a paired sample t-test on the 20 facilities (Managing Change mean 1995 = 54, mean 1997 = 52, df = 20, $t = -0.92$, $p < 0.37$).

Patient satisfaction

The overall satisfaction with quality of care and service delivery for inpatients ranged from 92 percent to 99 percent satisfied in 1997 (mean of 95 percent), and from 90 percent to 100 percent in 1995 (mean of 95 percent).

Table 4.2   Participation rates

| Facility code | n 1997 | n 1995 | Participation rate (%) | |
			1997	1995
A	189	181	76	76
B	372	286	78	50
C	510		99	
D	503		81	
E	269	187	100	56
F	497	441	80	77
G	217	202	84	54
H	127	58	86	25
I	196	130	88	70
J	228		78	
K	127	53	86	49
L	189	180	97	82
M	150	95	93	44
N	766		91	
O	278	195	82	63
P	241	173	100	63
Q	73	98	80	78
R	171	141	92	74
S	555	434	91	53
T	317	293	98	78
U	259	289	92	76
V	444		94	
W	56	55	98	52
X	85	79	99	60
Y	225	163	97	53
Z	263	167	95	74
	7307	3900	90	61

n = the number of returned scan sheets. The sample population from any one facility may be smaller than the return rate because some sheets were eliminated due to incomplete responses or high distortion scores.

Table 4.3   Descriptive statistics for managing change, patient satisfaction, and employee turnover

Variable	n	Mean	Standard deviation
Managing change 1997	26	50.8	8.65
Managing change 1995	21	53.7	8.78
Patient satisfaction 1997	25	0.95	0.02
Patient satisfaction 1995	17	0.95	0.02
Turnover 1997	25	29.6	15.81

These are very high overall levels of patient satisfaction. There was no significant difference between satisfaction levels in 1995 and 1997 (16 paired facilities mean patient satisfaction 1995 = 95 percent, mean 1997 = 96 percent, df = 15, t = 1.83, $p < 0.09$).

In addition to the measure of patient satisfaction taken from patients themselves, the Human Factors Inventory has one item asking about customer satisfaction from an employee's perspective: "Do you think your department meets the needs of its customers (e.g. patients, physicians, visitors, other departments, etc.)?" Although only scale score data were retained from 1995 (Managing Change scale), item level results by facility were still available for 1997. In 1997, 72 percent of employees on average said they felt they met the needs of their customers (very often/always and often; see Table 4.4). Responses from patients and from employees were highly convergent. The correlation coefficient for patient satisfaction and for employees reporting that they satisfied their customers was 0.42.

Interestingly, patient satisfaction can be differentiated from opinions of quality. Another Human Factors Inventory item asks employees: "Do employees in your department really care about quality?" This item was not related to patient satisfaction levels (r = 0.08, see Table 4.4). Thus, although quality is an important organizational variable, in the present study employees' opinions of caring about quality did not relate to satisfied patients. On the other hand, employee reports of satisfying their customers ("department meets the needs of its customers") did correlate with patients' reports of being satisfied with the service and care.

Employee turnover

Objective turnover statistics were available for 25 of the facilities for the fiscal year 1997. Turnover rates ranged from 9 percent to 80 percent with an average turnover rate of 29.6 percent. In addition to this objective measure, two items on the Human Factors Inventory address employee

Table 4.4 Patient satisfaction from the patients' view and from employees' view

Variable	n	Mean	Standard deviation	1	2
1 Patient satisfaction 1997 (rated by patients; % favorable)	25	95	0.02		
2 Department satisfies its customers (rated by employees; % favorable)	26	72	6.3	0.42* 25	
3 Department cares about quality (rated by employees; % favorable)	26	78	4.9	0.08 25	0.49* 25

Pearson correlation *$p < 0.05$, one-tailed/n

turnover. One item asks employees whether, "Turnover is high in your department." This item was rated on a strongly agree to strongly disagree Likert-type five-point scale. Forty-three percent of employees responded unfavorably (strongly agree or agree) to this item (see Table 4.5). The second turnover related item asks, "Do you think about quitting your job?" This item was also rated on a five-point Likert-type scale ranging from very often/always to never. Twenty-two percent of employees on average responded unfavorably (very often/always or often) to this item (see Table 4.5).

Interestingly, these two subjective measures of employee turnover differentially correlated with the objective measure of turnover. Table 4.5 shows that the Pearson correlation coefficient for 1997 rates of turnover and employee perceptions that turnover in their department was high is 0.67. Reported turnover intentions, however, were not significantly related to actual turnover rates ($r = 0.24$). This differentiation is not surprising. Conceptually, ratings of actual turnover ("turnover is high in my department") are more proximal to actual turnover rates than is intention to turn over ("I often think about quitting my job") (Steers and Mowday 1981).

Correlations between managing change and organizational outcomes

Table 4.6 shows the zero-order correlations among Managing Change (employee perceptions of managing change efforts in 1995 and 1997), satisfaction measures (patient rated satisfaction in 1995 and 1997, employee perceptions of satisfying customers in 1997, and employee perceptions of quality in 1997), and employee turnover measures for 1997 (actual turnover, employee perceptions of department turnover, and employee turnover intentions). All correlations were in the expected direction.

Table 4.6 indicates that the most strongly correlated measures were year-dependent. That is, Managing Change in 1997 was more strongly associated with patient satisfaction in 1997 than with patient satisfaction

Table 4.5 Objective and subjective measures of employee turnover

Variable	n	Mean	Standard deviation	1	2
1 Turnover in 1997 (%)	25	29.6	15.81		
2 Turnover is high in your department (% unfavorable)	26	43.2	0.09	0.67** 25	
3 Think about quitting your job (% unfavorable)	26	21.7	0.05	0.24 25	0.51** 26

*p<0.05; ** p< 0.001, one-tailed

Table 4.6 Correlations among managing change, patient satisfaction, and employee turnover

Pearson correlation n	1	2	3	4	5	6	7	8	9
1 Managing change 1997	—								
2 Managing change 1995	0.46* 21	—							
3 Patient satisfaction 1997	-0.45** 25	-0.31 19	—						
4 Department satisfies customers 1997	-0.52** 26	-0.12 21	0.42* 25	—					
5 Department cares about quality 1997	0.06 26	0.37* 21	0.08 25	0.49** 26	—				
6 Patient satisfaction 1995	-0.25 17	-0.42* 17	0.42 16	0.32 17	0.23 17	—			
7 Employee turnover 1997	0.40* 25	0.19 21	-0.60** 25	-0.41* 25	0.04 25	-0.24 16	—		
8 Turnover in department is high 1997	0.45* 26	0.21 21	-0.53** 25	-0.60** 26	-0.18 26	0.21 17	0.67** 25	—	
9 Think about quitting your job 1997	0.67** 26	0.21 21	-0.14 25	-0.57** 26	-0.24 26	-0.39 17	0.24 25	0.51** 26	—

$*p < 0.05$; $**p < 0.001$, one-tailed

in 1995. Likewise, Managing Change in 1995 was more strongly associated with patient satisfaction in 1995 than in 1997. The same pattern held true for employee turnover. Employee turnover in 1997 was more strongly related to Managing Change and patient satisfaction in 1997 than it was for these two factors measured in 1995.

Thus, Managing Change was related to the important organizational outcomes of patient satisfaction and employee turnover. With respect to patient satisfaction, facilities with poorer Managing Change efforts (higher Managing Change scores) had lower patient satisfaction ratings from both the patient and employee perspective. Patients reported higher levels of satisfaction with the care and services they received in the facility (1995 and 1997) and employees reported that their department satisfied its customers at facilities where employees also felt that the organization prepares well for the future, informs them about changes, communicates the reason behind the change, and helps them adjust to change.

Turnover data available in 1997 included not only objective turnover statistics, but also employee reports of turnover in their department and whether or not they were personally thinking about quitting their job. All three measures of turnover were significantly related to Managing Change efforts in 1997. Thus, employees who reported more effective Managing Change efforts also reported lower levels of department turnover and less frequent thoughts about quitting their job, and had lower levels of actual facility turnover than did their counterparts.

With respect to patient satisfaction, recall that there were two years (1995 and 1997) of patient satisfaction data provided by patients. Employees provided opinions of customer satisfaction in 1997 ("my department meets the needs of its customers"). Patients were more satisfied in facilities that had more favorable managing change scores (for 1995 and 1997). Likewise, patient satisfaction was greater in facilities with higher reports of customer satisfaction. Patient satisfaction was also related to actual facility turnover rates, as well as subjective rates of turnover in departments.

DISCUSSION

The purpose of this study was to investigate employees' perceptions of managing change efforts and determine whether efforts are related to important organizational outcomes. A large multi-facility hospital system was selected because of its corporate mission of facility acquisition, thus resulting in a constant state of change and growth for this corporation, as well as because of the inherent changes within the healthcare industry itself.

The findings support the importance of effectively managing the human-side of change in organizations. Employee perceptions of managing change were associated with the important organizational outcomes of patient satisfaction and employee turnover. Subjective measures confirmed the objective findings. Facilities that were rated as more effective at managing change had higher patient satisfaction and lower levels of employee turnover.

There are many limitations to the current study. First, objective measures of change efforts within facilities were not taken. All facilities implemented the "Speakeasy" program to promote effective communication. However, very little is known about other initiatives that were taken to assist employees in managing change efforts. We also know very little about how the Speakeasy program or other initiatives and activities impacted employee perceptions. Little is known about why some facilities improved, worsened, or stayed the same with respect to employee perceptions of managing change from 1995 to 1997.

Second, this case study was cross-sectional as opposed to longitudinal in design. It would have been advantageous if the 1997 employee responses could have been matched to their 1995 survey. In this way we could have observed "new" versus tenured employee perceptions and tenured employees over time. Measures of patient satisfaction, however, did follow employee perception measures. Survey administrations occurred in April and May, while patient satisfaction ratings were cumulated for July through September of the corresponding year.

Third, this study did not account for acquisition date and track pre- versus post-acquisition perceptions of managing change. Since the survey process was part of a larger risk management initiative, all facilities participated in the survey at the same time. On average, managing change scores were at or slightly above the 50th percentile, indicating average levels of managing change efforts; however, it is not known what level of actual change coincided with these perceptions. Some employees perceived their facility as extremely effective at preparing for the future and managing change, while others perceived their facility to be extremely poor at this type of management. One facility was in the top five worst facilities at managing change for both 1995 and 1997. This facility however, improved from the 80th to the 65th percentile. What efforts led to this improvement in 1997 and what is needed for continued improvement is not yet known. Also, we do not know what initiatives led employees in another facility to perceive managing change efforts so effective that their facility was the most favorably rated compared to all other facilities on managing change in both 1995 and 1997.

In continued efforts to understand these issues, facility risk managers and human resource employees are conducting department-level focus groups. We hope to identify best practices and lessons learned to enhance future management change efforts.

Although there are many limitations to this study, it is a first step in documenting the criticality of involving employees in an organization's change efforts. This study's findings support the fact that if your strategic efforts focus on financial figures at the exclusion of your workforce, for example if you do not share your vision for the future or you fail to inform employees of changes that affect them, you will not realize your organization's goals. Change is here to stay. An organization's ability to manage (not control) this change will determine its success.

References

Blecher, M. B. (1998) Wall Street smart, Main Street savvy, *Hospitals and Health Networks*, September 5, 1998, 38–42.

Bogner, M. S. (1994) (ed.) *Human Error in Medicine*, Hillsdale, NJ: Lawrence Erlbaum Associates.

Bond, M. (1997) At St. Paul, MN Hospital: 'Good Menu' delivers satisfaction increases, *Foodservice Director* 10(9): 76.

Finkel, M. L. (1997) The importance of measuring patient satisfaction, *Employee Benefits Journal* 22(1): 12–15.

Health Management Associates, Inc. (1998a) Health Management Associates, Inc. Fact Sheet, Summer 1998. Naples, FL: Health Management Associates, Inc.

Health Management Associates, Inc. (1998b) Health Management Associates, Inc. Reports its Fortieth Consecutive Quarter of Uninterrupted Earnings Growth, Press Release, October 20, 1998.

Helmreich, R. L. and Schaefer, H. (1994) Team performance in the operating room, in M. S. Bogner (ed.) *Human Error in Medicine*, Hillsdale, NJ: Lawrence Erlbaum Associates.

Larkin, H. (1997) California medical groups being ranked on quality, *American Medical News* 40(45): 5–6.

Mowday, R. T., Porter, L. W., and Steers, R. M. (1982) *Employee–Organizational Linkages: The Psychology of Commitment, Absenteeism, and Turnover*, New York: Academic Press.

Schneider, B. (1991) Service quality and profits: can you have your cake and eat it, too? *Human Resource Planning* 14(2): 151–7.

Sers, C. (1998) Is remaking the hospital making money? *Hospitals and Health Networks* July 20, 1998, 32–3.

Steers, R. M. and Mowday, R. T. (1981) Employee turnover and postdecision accommodation processes, in L. Cummings and B. Staw (eds) *Research in Organizational Behavior* 3, Greenwich, CT: JAI Press.

St Paul Fire and Marine Insurance Company (1995) *Human Factors Survey Process Handbook*, St Paul, MN.

Wiley, J. (1996) Linking survey results to customer satisfaction and business performance, in A. I. Kraut (ed.) *Organizational Surveys: Tools for Assessment and Change*, San Francisco, CA: Jossey-Bass.

Zimmerman, D. and Zimmerman, P. (1997) Customer service: the new battlefield for market share, *Healthcare Financial Management* 51(10): 51–3.

Chapter 5

Improving communications and health in a government department

S. Cartwright, C. L. Cooper, and L. Whatmore

INTRODUCTION

Occupational stress research has consistently identified poor communication as a potential source of stress and dissatisfaction in the workplace, particularly in a climate of uncertainty and change (Cooper and Roden 1985; McHugh 1995a; Nelson *et al.* 1995). In the context of major change events, such as corporate mergers and acquisitions, research studies have linked adequacy of merger-related communication with employee perceptions of the honesty and trustworthiness of the organization, levels of uncertainty and job satisfaction, as well as decisions to quit the newly formed organization (Cartwright and Cooper 1996; Schweiger and DeNisi 1991).

In more general terms, there is a well-established literature that has long argued that organizations should place a greater emphasis on upward and lateral communication processes as a means of improving their overall effectiveness. Research in this field has again continued to demonstrate the positive impact of employee involvement and consultation on a range of organizational outcomes, e.g. quality control (Guzzo *et al.* 1985) and increased productivity through participative goal setting (Locke and Latham, 1984). Indeed, the success of many large-scale business improvement initiatives like Total Quality Management (TQM) and Business Process Re-engineering (BPR) appears to be highly dependent upon the implementation of improved communication strategies and processes within an organization, particularly those geared towards greater team-working.

Communication is such a broad, multi-faceted and potentially overwhelming issue with organizations, that it is often difficult for companies to operationalize an espoused need and commitment "to improve communication" into a practical systematic strategy, which impacts at all organizational levels and is holistic in its approach. Examples drawn from the stress literature (Cooper *et al.* 1996) suggest that organizational efforts to improve communication in response to some form of stress audit are more likely to focus on certain employee groups and/or address selective aspects of communication. Typically, interpersonal skills training forms a major element of such intervention strategies.

The case study which forms the basis of this chapter discusses and evaluates the impact of a wide-ranging, multi-foci program of actions designed to improve communications, both formal and informal, and increase employee involvement in a large and major UK government department. Despite recent and deliberate efforts to move towards an "empowered" culture, public sector organizations, as traditionally command and control structures, have been highly resistant, and indeed positively discouraging in their attitudes towards the concepts of employee involvement and consultation (McHugh 1995b). The intervention was introduced following the findings of an initial stress audit, which formed the base line data against which the intervention was subsequently evaluated and its impact on employee health and well-being assessed. Results were also compared against a control group. The department already had a long-established in-house counseling facility, and so the intervention was directed at tackling the problem of stress at the primary and secondary level (Murphy 1988).

BACKGROUND TO THE STUDY

The government department concerned employs over 25,000 people across the UK. Sites vary in size from single individuals operating from remote locations to large city center offices with 1000+ employees. It is a diverse organization in terms of the work it performs. Its roles and responsibilities include law enforcement, tax revenue collection and the compilation of statistical information. It is also diverse in terms of the background and educational qualifications of the people it employs. There are a significant number of shift workers and a small, but increasing number of contract workers. Like most government departments, it has and continues to undergo major change. In common with other public sector organizations, it has problems of escalating sickness absence and stress-related illness well above the UK average.

In recent years, the department has experienced radical changes in its policies, practices, culture and style of work organization including the introduction of service level agreements and performance targets. It has invested substantial resources and efforts in trying to move the culture of the organization in a direction which empowers people to make their own decisions and encourages individuals to work as teams. However, there are major tensions between the new espoused or desired culture and the traditional grade-conscious culture of the department, which stifles challenge and criticism and rewards and promotes technical as opposed to "people" management skills. Communication patterns within the department still reflect its hierarchical structure and are predominantly downward.

In 1996, as part of a wider initiative to address the problem of rising sickness absence rates across the department, a stress audit was conducted. This involved the collection of qualitative (interviews and discussion groups) and quantitative (questionnaire) data from two corporate divisions located in the south-east of England. The division which formed the experimental group had approximately 530 employees, whereas the division which acted as the control group had just over 200 employees.

The initial stress audit procedure

A general communication was separately put out to both divisions inviting employees to participate in the initial study by attending either an individual interview (higher managerial grades) or focus group sessions (all other employee grades), to discuss the factors which contribute to workplace health and well-being. In total, 6 individual interviews and 12 focus group sessions were conducted involving 105 employees. Care was taken to ensure the absence of any potential boss/subordinate relationships within the groups.

Both interviews and discussion groups followed a semi-structured format based around the stress research model (Cooper and Marshall 1978) from which the Occupational Stress Indicator (OSI) is derived. Participants were asked to discuss the issues as openly and honestly as they could, and assured that all individual details would remain confidential to the researcher. They were also encouraged to express any concerns or issues that they considered were representative of their workgroup colleagues. Several participants had taken the opportunity to discuss issues relating to stress with their co-workers prior to attending the sessions. It was explained that notes would be taken, summarized and fed back to participants at the end of the session to ensure their accuracy.

Two weeks later, questionnaires, comprising the OSI and a biographical questionnaire, were sent to all employees in both the experimental and control condition. The OSI (Cooper *et al.* 1988) is an integrated self-report measure of sources of stress, Type A behavior, locus of control, job satisfaction, employee health and coping skills, arranged on seven scales. Normative data are available, drawn from a sample of over 14,000 mainly white-collar workers. Employees completed the questionnaires anonymously and returned them direct to the university.

Of the 540 questionnaires distributed in the experimental division, 253 were returned by the cut-off date, representing a 47 percent response rate. The response rate from the control division was lower but not that dissimilar. Of the 226 questionnaires distributed, 90 were returned, representing a response rate of 40 percent. Information from the biographical section of the questionnaire indicated that the sample was representative of the total population in terms of age, gender, grade, etc.

FINDINGS

Stress levels

Compared to normative data, levels of stress in both divisions were significantly higher than the general working population (Table 5.1) on five out of the six subscales ($*p < 0.001$), the exception being factors related to the Home/Work interface. Furthermore, statistical analysis indicated that there were no significant differences between the two divisions on any of the subscales.

Individual questionnaire item analysis and interview/discussion group data were used to expand upon the nature of the stressors operating in the organization. It can only be very briefly summarized here.

The job itself

Stress emanating from the job itself centered around high work volume and tight deadlines aggravated by managers not delegating work quickly enough.

Role in the organization

One of the main issues concerned the expectations of management, which were perceived as being both unclear and changing.

Relationships with others

Data analysis indicated that the main issue was relationships with superiors, their poor people management skills and a continuance of a "blame" culture in many areas.

Table 5.1 Initial stress audit – sources of stress

Stressor	Experimental group (n = 253)		Control group (n = 90)		Norms (n = 14,455)	
	Mean	SD	Mean	SD	Mean	SD
Job itself	**30.65	6.70	*30.77	6.97	29.28	6.35
Role in the organization	**35.73	9.02	35.43	8.71	33.99	8.21
Relationships with others	**32.55	7.75	*31.52	7.76	29.84	7.37
Career and achievement	**32.87	7.47	**32.17	7.87	28.87	7.73
Org. structure and climate	**41.98	9.14	**40.68	8.93	37.47	8.88
Home/work interface	30.35	12.04	30.96	10.40	30.45	3.56

*p < 0.05; **p < 0.001

Career development and achievement

Stress emanating from this source fell into three main worries: job insecurity, the lack of opportunity for career development, and the appraisal system. In terms of percentage difference from norms, this category of stressor was the greatest source of stress for employees.

Organizational structure and climate

In percentage terms, stress emanating from the organizational structure and climate came a close second to Career and Achievement and was strongly related to inadequate communication and lack of consultation in the change process. A common complaint was that there were too many changes in too short a time with an apparent lack of direction, resulting in confusion and uncertainty amongst staff.

Employees felt that too much reliance was made on providing information through weekly bulletins and newsletters, which often they did not have time to read. Face-to-face communication was reported to be virtually non-existent. The rationale for decisions was rarely or minimally communicated and resulted in many rumors and mis-information.

Job satisfaction and employee health

The total job satisfaction mean scores for both divisions, Experimental ($x = 72.88$ SD 14.70) and Control ($x = 70.46$ SD 15.34) were significantly below the norm ($x = 82.15$ SD 15.94; $p < 0.001$). Similarly, physical and psychological healthcare scores (Table 5.2) were comparable between divisions and significantly poorer ($p < 0.001$) than normative (general population) data.

Moderators of the stress response

Table 5.3 shows the mean scores and standard deviations for the total Type A scale and total locus of control scale.

Table 5.2 Physical and mental health – pre-intervention

	Experimental group (n = 253)		Control group (n = 90)		Norms (n = 14,455)	
	Mean	SD	Mean	SD	Mean	SD
Mental health	**58.99	14.38	**59.67	15.97	52.98	13.81
Physical health	**35.58	10.66	**34.50	10.66	30.64	9.98

**$p < 0.001$; high score = poor health

Table 5.3   Type A and locus of control scores – pre-intervention

	Experimental group (n = 253)		Control group (n = 90)		Norms (n = 14,455)	
	Mean	SD	Mean	SD	Mean	SD
Total Type A	46.86**	6.75	45.91**	6.75	50.12	7.38
Total locus of control	44.30**	5.46	44.06**	5.89	41.85	5.32

**$p < 0.001$; high score = high Type A and externality

In terms of these characteristics, the populations of the two divisions are comparable and are significantly less Type A ($p < 0.001$) and perceive themselves to be more externally controlled ($p < 0.001$) than the general population. Table 5.4 shows the mean scores for the two divisions for the coping scales.

The intervention

The results of the stress audit were of concern to the organization when they were presented at board level. Given the many and varied categories of stressors identified by the audit, the board was presented with a potentially overwhelming number of possible areas at which to target interventions, but with understandably insufficient resources to tackle them all. After much deliberation, responsibility for the design, focus and introduction of any subsequent intervention strategy was assigned directly to the recently appointed Head of Division (HOD). Because the initiative was at a local rather than a national departmental level, this effectively restricted the foci and types of intervention that practically could be introduced.

Table 5.4   Use of coping strategies – pre-intervention

Strategy	Experimental group (n = 253)		Control group (n = 90)		Norms (n = 14,455)	
	Mean	SD	Mean	SD	Mean	SD
Social support	15.50	3.17	15.00	3.05	15.46	20.4
Task management	**25.09	3.51	25.91	3.10	25.73	3.63
Detached response (logic)	**11.89	2.13	12.43	1.89	12.30	2.08
Home/work relationships	*16.13	3.19	15.79	3.63	15.58	3.43
Time management	14.24	2.08	14.51	1.83	14.33	2.17
Involvement	23.30	3.42	22.91	3.22	23.12	3.31

*$p < 0.05$; **$p < 0.01$; high score = high use of strategy

Whilst it was clear from the stress audit that a primary level intervention designed to eliminate or modify stressors associated with "career and achievement" would be an ideal "first choice" action, this would require changes in employment policies and the appraisal system which could not be determined and implemented locally. Consequently, the HOD decided to concentrate on the second ranked cluster of stressors relating to "communication" and the "organizational structure, climate and culture." Part of the strategy developed involved an operational review of a number of policies, systems and procedures including the appraisal process. These reviews were carried out by working parties in consultation with employees at all levels in the organization, the rationale being that whilst the division could not change the system, improvements could be made at an operational level to ensure that it was interpreted and implemented in a fair and consistent way across the division.

Levels of communication and foci of interventions

The basic three-box model of communication conceptualizes communication as a process involving three constituent parts; the sender, the message and the receiver (Schermerhorn *et al.* 1982). Effective communication is communication in which the intended meaning of the sender or source and the perceived meaning of the individual or groups of individuals who receive the message are one and the same. Barriers to effective communication are many and various, but broadly can be categorized as:

- factors relating to the sender, e.g. perceived credibility and attractiveness of the source, communication style, etc. (Gross 1987)
- factors relating to the message e.g. language, structure and content, the appropriateness and effectiveness of the media chosen to deliver the message etc. (Lengal and Daft 1988)
- factors relating to the receiver, e.g. attitudinal disposition and emotionality of the recipient, feedback mechanisms, etc. (Hargie 1987)

In the context of improving communication, the division introduced 22 separate initiatives as part of its overall strategy (see Table 5.5).

The strategy incorporated significant changes that addressed all three aspects of the communication process described above. The basic tenets of the strategy were:

1   Increased visibility and direct communication from the HOD. The HOD introduced a series of small group Question and Answer sessions at which employees had the opportunity to question him directly and gave him the chance to fully explain and clarify the rationale behind organizational decisions and the future direction of the business.

Table 5.5   The 22 separate initiatives

Procedural and systems reviews
    Newcomers/induction course
    Communication
    Appraisal
    Training and development needs
    Bonus scheme

Organizational development initiatives
    Delayering and empowerment
    Continuous improvement events
    Clarification of roles
    Development events
    Work associated with Investor in People accreditation
    Publication of strategy document

Skills training
    Stress management and awareness
    Refresher courses on appraisal system
    Team-building events
    Leadership development program

Employee-centered initiatives
    Introduction of long service awards
    Bright ideas scheme
    Q and A sessions
    Interdepartmental presentations/open days
    Charity events
    Dissemination of findings of stress audit/stress information sheets
    Letting-off-steam initiatives, e.g. scream boards

2   A widening of the media used to deliver information through increased emphasis on face-to-face communication and informal communication networks. Part of the strategy involved the introduction of a program of social and charity-raising events to encourage managers and employees to get to know each other outside a work context.

3   Increased opportunities for employee involvement and feedback. In addition to the systems and procedural reviews, a "bright ideas" scheme was set up whereby employees were encouraged to submit their suggestions for improvement and in return were rewarded with a novelty mug. Upward and lateral communication was promoted by initiatives such as the "scream" board and a series of inter-departmental presentations.

"Scream" boards were introduced in a number of teams, whereby team members could anonymously record specific instances and issues that occurred throughout the course of the week, and which they considered caused them stress. At the end of the week, the team leader discussed the recorded items with their teams and the possible actions that could be taken to alleviate these pressures. The purpose of the inter-departmental presentations was to inform others of the role and nature of the work undertaken by different departments to improve organizational coherence and promote a greater understanding of the internal supplier/customer relationship.

In the context of stress reduction, an alternative model of conceptualizing communication in organization is being at four interacting levels; individual (intra-communication), interpersonal, group/departmental and organizational (model 1).

This model proposes that the destructive/constructive nature of internal mental dialog affects the way in which the individual appraises organizational events and experiences, and influences their behavior and the way in which they communicate and respond to others. At a collective level, this in turn will shape the communication patterns and culture within an organization, and so continue to reinforce the attitudes and beliefs of its individual members. If individuals believe that they are unable to exert influence and have little or no control over events around them, the resultant feelings of powerlessness and learned helplessness lead to experienced stress and inhibit upward communication.

The findings of the stress audit indicated high "externality" in terms of control amongst employees reinforced by the organizational structure, climate and communication processes. As 86 percent of the sample had been in the department for more than six years and almost half of the

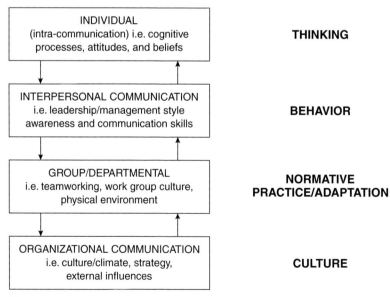

Model 1

entire population had over 15 years of service, any attempts to improve communication also needed to address the strongly established internalized beliefs of employees as well as the three, more obvious, levels of the model.

Therefore, as part of this holistic, multi-level approach, a major part of the stress awareness/stress management training provided centered around rational-emotive therapy (RET) (Ellis 1962) and constructive self-talk, to encourage employees to challenge the irrationality of their beliefs, make more positive appraisals of events and restore their perceptions of internal control. As a means of stress reduction, cognitive techniques have been shown to be less effective as an individual strategy than exercise in a similar organizational setting (Whatmore *et al.* 1998). However, such training was not complemented by any reinforcing organizational action to support and maintain any changes in cognition. Although the training was introduced on a voluntary basis, about 30 percent of the total population of the division participated in this initiative.

Post-intervention evaluation

Procedure

Although some initiatives (e.g. the communications review; delayering and empowerment and team-building events) were introduced in 1996 and are still ongoing, the majority came into action during the period September 1997 to March 1998. Some initiatives were single one-off events (e.g. the publication of the strategy document), others are ongoing and have become part of the culture.

The re-audit was conducted in spring/early summer 1998 in both the experimental and control divisions. At the time, the experimental division was about to reorganize which was likely to result in jobs being downgraded.

The procedure was similar to the initial audit and involved a questionnaire and focus group discussions. Focus group discussions were concerned with establishing current stressors and changes since the first audit. In the experimental condition, they provided the opportunity to discuss the effectiveness and helpfulness of the interventions introduced. About half of those who attended the focus groups had attended the initial sessions.

Of the 540 questionnaires distributed in the experimental division, 245 returned questionnaires, representing a response rate of 45 percent. Biographical information, attitudes towards and participation in the 22 interventions were collected; 103 (61 percent) individuals reported that they had participated in the previous audit, 37 (22 percent) had not, and 28 (17 percent) could not remember. Again, the sample was representative of the total population in terms of age, gender, grade, etc. Although, to ensure anonymity, it was not possible technically to match data across the two audits, no significant differences were found between first- and second-time respondents, and as far as possible we can be sufficiently confident of the validity of the comparative data. Out of the 225 questionnaires dispatched to the Control Division 93 were returned, representing a response rate of 41 percent.

FINDINGS

Stress levels

Levels of stress in the experimental division were similar to those at the time of the first audit with the exception of stress emanating from the Organizational Structure and Climate which was found to be significantly lower ($p < 0.05$) post-intervention. Stress levels in the control group remain unchanged (see Table 5.6).

Job satisfaction and employee health

There were significant pre- and post-intervention improvements in job satisfaction in the experimental division ($p < 0.001$) whereas there was no change in the control division. These improvements impacted on all the job satisfaction subscales. However, there were no significant changes in physical and psychological health (see Table 5.7).

Moderators of the stress response

Table 5.8 shows the mean scores and standard deviations for the total Type A scale and total locus of control scale for the two divisions.

Table 5.6   Stress re-audit: sources of stress

Stressor	Experimental group				Control group (n = 93)	
	Pre-intervention (n = 253)		Post-intervention (n = 245)			
	Mean	SD	Mean	SD	Mean	SD
Job itself	30.65	6.70	30.69	6.13	30.92	6.18
Role in the organization	35.73	9.02	36.55	8.20	36.59	8.71
Relationships with others	32.55	7.75	32.79	7.34	32.59	7.94
Career and achievement	32.87	7.47	32.32	7.28	31.56	8.07
Org. structure and climate	41.98	9.14	40.18*	8.78	41.67	8.69
Home/work interface	30.35	12.04	31.12	11.28	32.67	11.27

*p significant @ < 0.05

Table 5.7 Job satisfaction and employee health – post-intervention

	Experimental group				Control group (n = 93)	
	Pre-intervention (n = 253)		Post-intervention (n = 245)			
	Mean	SD	Mean	SD	Mean	SD
Total job satisfaction	72.88	14.70	78.22*	16.29	74.81	15.37
Mental health	57.96	14.70	58.99	14.38	60.00	13.80
Physical health	35.96	35.88	10.99	10.66	35.44	9.34

*p significant @ < 0.001 (pre/post)

Table 5.8 Type A and locus of control scores – post-intervention

	Experimental group				Control group (n = 93)	
	Pre-intervention (n = 253)		Post-intervention (n = 245)			
	Mean	SD	Mean	SD	Mean	SD
Total Type A	46.86	6.75	46.59	6.97	46.57	6.41
Total locus of control	44.30	5.46	42.43**	5.36	44.02*	5.65

*p significant @ 0.5 (post/control); **p significant @ < 0.001 (pre/post)

Although there was no significant change in Type A behavior post-intervention, there was a significant change in locus of control (LOC) scores. Comparative to pre-intervention data employees in the experimental division felt that they had significantly more control and influence over events, i.e. are more internal, than they did at the time of the first audit ($p < 0.001$).

There was also an observed change/improvement in the use of coping strategies pre- and post-intervention in the experimental group (Table 5.9). In terms of the use of task management, this difference achieved statistical significance ($p < 0.001$).

Employee attitudes

Focus group data indicated that employees in the experimental division found it to be a more positive place to work and that the interventions had changed the culture. Management was perceived to be more approachable and engaged in more face-to-face communication. There was also reported

Table 5.9   Use of coping strategies – post-intervention

Strategy	Experimental group				Control group (n = 93)	
	Pre-intervention (n = 253)		Post-intervention (n = 245)			
	Mean	SD	Mean	SD	Mean	SD
Social support	15.50	3.17	15.77	3.17	15.18	3.11
Task management	25.09	3.51	26.18**	3.44	25.64	3.03
Detached response (Logic)	11.89	2.13	12.17	1.86	11.95	2.04
Home/work relationships	16.13	3.19	16.27	3.49	16.03	3.51
Time management	14.24	2.08	14.50	1.92	14.47	1.90
Involvement	23.30	3.42	23.82*	3.04	22.67	2.97

*p significant @ $p < 0.01$ (post/control); **p significant @ $p < 0.001$ (pre/post)

to be better inter-departmental working, more discussion and cooperation across teams and that generally people were talking more to each other. Stress events had been particularly useful in providing a common language and making it more acceptable to talk about stress. The improvements in communication were widely recognized. In terms of stress reduction, the inter-departmental presentations (information giving open days), the Question and Answer sessions and the review of the induction course were rated most highly. The majority of employees considered that all of the 22 initiatives had been helpful in reducing stress and improving communication. This was confirmed by an internal attitude survey conducted by the division a few months before the audit which found that 80 percent of respondents were satisfied with organizational communication.

Conclusion

The evidence from this case study suggests that improvements in communication positively impact on job satisfaction and employees' perceptions of control and influence. In view of the positive relationship between "internal" locus of control and ability to cope with stress, this is encouraging. The intervention was primarily designed to address stress emanating from the "organizational structure and climate," and in this respect it was successful in that stress levels relating to this factor have reduced significantly. However, stress levels within the experimental division still compare unfavorably with normative data and there has been no measurable impact on health. It may be that increased job satisfaction may be a precursor to improved health, and as job satisfaction may in itself affect health symptoms, this may result in beneficial health outcomes in the longer term.

Within the context of public sector organizations, the type and breadth of this intervention is extremely radical and innovative. Whilst the self-report health measures were indicative of little change in physical and psychological health in the experimental division, there have been changes in sickness/absence patterns. At the time of the initial stress audit, absence due to mental disorders accounted for 10 percent of total days lost in both divisions. This figure has remained consistent in the experimental division but has increased to 30 percent in the control divisions.

References

Cartwright, S. and Cooper, C. L. (1996) *Managing Mergers, Acquisitions and Strategic Alliances: Integrating People and Cultures*, Oxford: Butterworth-Heinemann.

Cooper, C. L. and Marshall, J. (1978) *Understanding Executive Stress*, London: Macmillan.

Cooper, C. L. and Roden, J. (1985) Mental health and satisfaction amongst tax officers, *Social Science and Medicine* 21(7): 474–81.

Cooper, C. L., Liukkonen, P. and Cartwright, S. (1996) *Stress Prevention in the Workplace: Assessing the Costs and Benefits to Organizations*, Luxembourg: European Foundation for the Improvement of Living and Working Conditions.

Cooper, C. L., Sloan, S. J. and Williams, S. (1988) *Occupational Stress Indicator Management Guide*, Windsor, UK: NFER Nelson.

Ellis, A. (1962) *Reason and Emotion in Psychotherapy*, New York: Lyle Stuart.

Gross, R. D. (1987) *Psychology: The Science of Mind and Behaviour*, London: Edward Arnold.

Guzzo, R. A., Jette, R. D. and Katzell, R. A. (1985) The effects of psychologically based intervention programs on worker productivity: A meta analysis, *Personnel Psychology* 38: 275–92.

Hargie, O. D. W. (1997) Interpersonal communication: a theoretical framework, in O. D. W. Hargie (ed.) *The Handbook of Communication Skills*, London and New York: Routledge.

Lengal, R. H. and Daft, R. L. (1988) The selection of communication media as an executive skill, *Academy of Management Executive* 2(3): 225–32.

Locke, E. A. and Latham, G. P. (1984) *Goal Setting: A Motivational Technique That Works*, Englewood Cliffs, NJ: Prentice Hall.

McHugh, M. (1995a) Organizational merger: a stressful challenge? *Review of Employment Topics* 3(1): 126–56. Belfast: Labour Relations Agency.

McHugh, M. (1995b) Stress and strategic change, in D. Hussey (ed.) *Rethinking Strategic Management*, Chichester: John Wiley.

Murphy, L. R. (1988) Workplace interventions for stress reduction and prevention, in C. L. Cooper and R. Payne (eds) *Causes, Coping and Consequences of Stress at Work*, Chichester and New York: John Wiley.

Nelson, A., Cooper, C. L. and Jackson, P. (1995) Uncertainty amidst change: The impact of privatisation on employee job satisfaction and well being, *Journal of Occupational and Organisational Psychology*, 68: 57–71.

Schermerhorn, J. J. Jr, Hunt, J. G. and Osborn, R. N. (1982) *Managing Organisational Behaviour*, New York and Chichester: John Wiley.

Schweiger, D. M. and DeNisi, A. (1991) Communication with employees following a merger: a longitudinal field experiment, *Academy of Management Journal* 34: 110–35.

Whatmore, L., Cartwright, S. and Cooper, C. L. (1998) Stress interventions in the UK: An evaluation of a stress management programme in the public sector, in M. Kompier and C. L. Cooper (eds) *Improving Work, Health and Productivity Through Stress Prevention: 14 European Cases*. London and New York: Routledge.

Chapter 6

The effects of promoting organizational health on worker well-being and organizational effectiveness in small and medium-sized enterprises

K. Lindström, K. Schrey, G. Ahonen, and S. Kaleva

INTRODUCTION

Organizational health implies that an organization is able to optimize its effectiveness and the well-being of its employees, and to cope effectively with both internal and external changes. A healthy organization is characterized by the good congruence of written values and rules with the everyday practices reflecting these values and rules (Cooper and Cartwright 1994; Cox and Leiter 1992). The healthy organization model has been linked in case studies to successful companies which also have a positive employee morale (Jaffe 1995; Levering 1988). This model has included personal job satisfaction related to the effectiveness of the organization. In a US manufacturing company Lim and Murphy (1997) found two pathways leading to organizational health. Organizational climate and values had an influence on organizational effectiveness, whereas organizational practices explained job satisfaction and stress.

It is still somewhat open to debate as to which organizational practice and climate factors are the most important contributors to organizational health. But in everyday practice it is especially important to know what kinds of individual and organizational interventions are successful in promoting organizational health. In development of the work organization, personnel-oriented approaches and the approaches to developing technology and business are still too far from each other, although sociotechnical job redesign practices do try to combine these. The promotion of healthy organization principles can be seen as a value-based strategy which combines business and human needs in an organization. In ideal cases the focus in a workplace intervention is both on individual employees and on the whole company or organization. The participatory approach has been favorable when aiming to improve organizational health. Especially when companies want to adapt and cope with increasingly complex and rapidly changing environments, the ideas of the learning organization are advantageous. When the company is able to mobilize the learning of the employees towards a process of continuous self-transformation (Starkey 1998), the balance between the company and its business environment can be mastered more successfully.

Larger companies or organizations are usually more active in implementing new forms of work organization, as well as planned organizational change and interventions. Small and medium-sized enterprises (SMEs) have not been as active in organizational development, usually because they lack the expertise, time and economic resources. In the present study, however, the European Social Fund financed the individual- and organization-oriented interventions among SMEs from Southern Finland (Huuskonen *et al.* 1997).

The first aim of the study was to determine how the job and organizational characteristics and the well-being of workers differed between branches of industry and size of the SMEs, and how the well-being of workers was related to job and organizational characteristics. The second and main aim was to study how the job and organizational characteristics and well-being of workers changed during a two-year follow-up, and how these changes were related to planned organizational interventions carried out. The final aim was to investigate how the planned interventions, job and organizational characteristics, and well-being of the employees were related to organizational effectiveness (i.e. productivity and profitability) of the SMEs.

STUDY GROUP

The project was started by contacting the SMEs in southern Finland through their occupational health services (OHSs). This contact channel was chosen because OHSs are mandatory in Finland for all employees, and OHS personnel are in continuous contact with the workplaces. The design of the data collection and interventions is illustrated in Figure 6.1. In total, 343 workplaces participated in the first round of the project in 1996, and 304 in the second round in 1997–98. These enterprises complied with the SME definition of the EU. In Round 1, questionnaires were distributed to all these workplaces through the OHSs in March and June 1996. In the winter of 1997–98, a second survey was administered to the same enterprises and their employees. This time, 217 employers and 4068 employees responded from among those who had participated in the first round. These companies and employees formed the study group described in this chapter. The distribution of the companies and employees by branch of industry, and the employees' sex distribution are shown in Table 6.1. At the second study round 39 enterprises dropped out from among those still participating in the project. The reasons for dropping out were: the actual situation

**March–June
1996**

**Winter
1997–98**

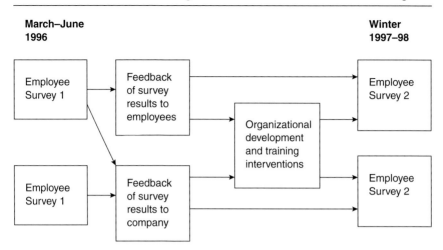

Figure 6.1 Measurement and intervention design of the study

Table 6.1 Number of employers and employees participating in both study rounds according to the branch of industry and gender

Branch	Number of enterprises	Employees		
		Total number	Men (%)	Women (%)
Food processing	8	181	25	75
Publishing and printing	15	302	41	59
Manufacture of electronic appliances	8	117	85	15
Motor vehicle trade and repair	11	206	82	18
Service stations	6	20	60	40
Hotels and restaurants	8	99	33	67
Metal and engineering industry	53	1408	81	19
Construction	42	586	85	15
Transport	14	299	85	15
Retail	19	374	26	74
Accounting and other office work	29	374	32	68
Other	4	102		
	217	4068	65	35

at the workplace was unfavorable, the OHS personnel had not time to organize the second survey, or there was bankruptcy or a merger of the enterprise.

Most of the enterprises in the study group were from the metal and engineering industry (24 percent) and construction (19 percent). Thirty-five percent of the participating employees were from the metal and engineering industry; 65 percent were men and 35 percent women. The high percentage of male workers was due to the selection of the enterprises in the sample, since manufacturing and traffic branches were more prevalent compared to service and office work.

Methods

Two questionnaire surveys were carried out, one for the employees and one for the employers or their representatives. These surveys were administered twice, before and after the planned interventions.

The employees' perception of their physical work environment and job and organizational characteristics as well as various aspects of well-being were measured by the survey. The physical work environment was described by the number of risk factors perceived as harmful (range 1–16). The job characteristics measured were physical workload (sum scale of three items, range 0–3), time pressure (a single item) and job control (scale of four items). Organizational practices and climate were measured using the following scales:

- supervisory support (1 item)
- relation to co-workers (1 item)
- continuous improvement practices (5 items)
- informing about changes (1 item)
- future insecurity of job (5 items)
- social climate (2 items)
- appreciation of one's work (1 item)

The well-being of the employees was based on their own reporting of psychological strain symptoms (7 items) and lack of psychological resources (3 items) which were combined to form a single measure of exhaustion. The respondents also evaluated job satisfaction with a single item question as well as their work capacity on a scale from 1 to 10. In addition, they reported how many days they had been away from work due to sickness during the past 12 months. Also lifestyle factors, such as problems with alcohol and participation in physical activities and sports, were asked.

At the first study round, needs for improvement in customer service, multi-skilling, leadership practices and collaboration were inquired from the employees and employers. The interventions or training carried out on these same topics were asked from both the employees and the employers at the second study round. Also the use of the participatory approach in interventions was asked from both groups.

The employers' survey covered their evaluation of the company's relative effectiveness as measured by the company's production of products and services and the company's profitability in 1996 and 1997. In these evaluations, they were asked to use 100 as indicating a satisfactory level of effectiveness. The employers or enterprise representatives also assessed the general level of work capacity of workers, and any changes in company structure, business, personnel and competition on the markets that may have occurred during the follow-up period.

Development needs and interventions carried out in the companies

At the first study round both the employees and employers reported about the development needs in company. The development needs that came up in the first survey are shown in Table 6.2. The employers saw needs in all the areas of development quite equally, and the employees reported most often development needs in leadership practices and collaboration.

The study model, including the interventions, is illustrated in Figure 6.2. All the enterprises and each employee received feedback on the results of the first survey. The individual results were compared to the reference data of the whole survey in order to allow benchmarking. The feedback also included recommendations for improvements and interventions. Based on the feedback data, and on the expressed needs for training and development in the first study, organizational or training interventions were planned and implemented in the different areas and to various extents at company level. The interventions were carried out by the workplaces themselves using the project team, OHS personnel, or outside experts as consultants.

Table 6.2   Development needs in 1996 and planned interventions at own workplace during the follow-up 1996–1997/98 reported by employees (n = 4068) and employers (n = 206)

	Type of intervention (at least to some extent %)			
	Customer service (%)	Multi-skilling of personnel (%)	Leadership practices (%)	Collaboration (%)
Development needs reported by				
employees	28	32	43	41
employers	37	38	40	*
Planned interventions reported by				
employees	54	40	31	41
employers	62	64	54	73

*not asked

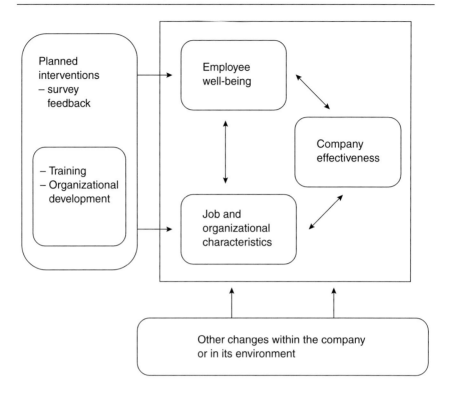

Figure 6.2   The intervention model of the study

Table 6.2 shows the percentage of employees at those workplaces where a planned intervention was carried out according to the employers' report. Most of the employers reported interventions aimed to improve collaboration within the company, and to improve multiskilling of workers. Half of the employers reported activities designed to improve management practices. These same questions were also answered by the employees, and their corresponding proportion of participation in the intervention was clearly lower than that of the employers. The employees reported most often needs in customer service development.

The employers also reported other essential changes that had taken place after the first survey within the company or in its business environment. Most of them mentioned some renewal in the work facilities or the work environment (20 percent), and changes in personnel (28 percent). Also investments in new machines and equipment (14 percent) had been made, as well as structural changes in the organization (12 percent). Increased competition in business was reported by 11 percent of the companies.

Statistical analyses

Comparisons were made between the employee survey data on the perceived working conditions, job and organizational characteristics, and well-being between the different-sized enterprises and between branches of industry, as well as between the first and second study rounds. In these comparisons the percentages of the subgroups were used and no statistical testing was applied. The relations between job and organizational characteristics and well-being were analyzed by Pearson's correlations.

The individual differences between the first and second study rounds in job and organizational characteristics and well-being were either correlated to the coverage of the planned interventions or were compared between the subgroups reporting more intensive and minor or no interventions. In this analysis, either Pearson's r correlations or Student's t-test were applied. Finally, stepwise multiple regression analysis was used to explain the changes in well-being by changes in job and organizational factors and the intensity of the interventions.

The two relative company-level effectiveness measures were related to the interventions carried out and to the employees' perception of their job and organizational characteristics and well-being in 1997, using Pearson's correlation. Also here stepwise multiple regression analysis was applied in order to explain the company effectiveness. In these analyses the employee data were aggregated to the enterprise level by summing the values of each employee from the same company and dividing the sum by the number of employees.

RESULTS

Job organizational characteristics and well-being and their interrelations in SMEs: First study round

Job and organizational characteristics and personnel well-being according to company size and branch of industry

In the first study round the job characteristics and organizational practices and climate differed among the companies of different size and belonging to different branches of industry. In particular, employees from enterprises with less than 10 employees had better job control, supervisory support, work climate and appreciation of one's own work. Also information concerning changes was perceived better in these smaller enterprises (Table 6.3).

Table 6.3 Enterprise size in relation to the job and organizational characteristics in the first and second study rounds (n = 4068)

| | Enterprise size (= number of employees) | | | | | | | |
| | 1–9 | | 10–49 | | 50– | | Total | |
	First round (%)	Second round (%)	First round (%)	Second round (%)	First round (%)	Second round (%)	First round (%)	Second round (%)
Physical work environment								
risk factors in work environment, 4 or more	32	24	26	22	31	20	29	21
Job characteristics								
physical work load, at least moderate	32	25	31	28	30	25	30	26
time pressure, continuous	20	27	24	33	22	30	23	31
job control, good or rather good	37	40	24	27	21	25	24	27
Organizational practices and climate								
supervisory support, always or often	48	46	40	36	39	38	40	38
co-worker relations, very or rather good	88	87	77	73	73	77	76	71
continuous improvement, always/nearly always	26	25	20	19	19	16	20	18
information about changes, always/nearly always	50	48	40	35	33	30	38	34
job insecurity, always or nearly always	25	16	28	16	28	17	28	16
positive social climate	64	67	49	44	41	37	46	42
appreciation of one's work, always/always	42	42	30	27	25	22	28	26

The employees from the various branches differed somewhat in their perception of job and organizational characteristics. Although various branches of the industry had somewhat different gender distributions (Table 6.1), it was mainly the different nature of the work which affected how the physical working conditions, job characteristics and organizational practices and climate were perceived. The highest time pressure was reported in accounting and other office work. The service stations were best in the perception of their organizational practices and climate, which may be due to the small size of the workplaces and younger age of workers. Supervisory support was perceived best in accounting and other office work, and in retail shops. The construction industry was a sector with good co-worker relations, but high insecurity as regards job future. Appreciation of one's own work was highest in service stations, and hotels and restaurants.

Of the well-being measures, only sickness absenteeism was related to the size of the company (Table 6.4). Those working in enterprises with over 50 employees had more sickness absenteeism days than those in smaller enterprises. Again the employees from various branches differed in well-being. Most exhaustion symptoms were found in accounting and other office work, and in motor vehicle repair and sales. The poorest working capacity was reported in transport, and in the manufacturing of electronic appliances; sickness absenteeism days were the highest in hotels and restaurants.

Interrelations of well-being measures, job and organizational factors, and some background factors

The interrelations between the job and organizational characteristics, well-being and some background factors were quite similar at both study rounds. Thus only the results of the first round are commented on here. The interrelations of four applied well-being measures were statistically significant, and the exhaustion symptoms correlated moderately with both low work capacity ($r = -0.37, p < 0.001$) and low job satisfaction ($r = -0.34, p < 0.001$).

The work environment factors and job and organizational characteristics were also correlated with each other (Pearson's r). The higher number of risks in the work environment and physical workload were clearly related to each other. Both time pressure and insecurity concerning job future had only rather low relations to any other job and organizational factors. Job control correlated moderately with appreciation of one's work, continuous improvement practices and efficient informing about changes. Co-worker relations were most clearly related to general work climate. Supervisory support, continuous improvement practices, informing about changes, social work climate, and appreciation were all rather highly inter-correlated. These probably reflected a workplace where supervisory support

Table 6.4 Well-being and company size in the first and second study rounds (n = 4068)

| | Company size, number of employees | | | | | | | |
| | 1–9 | | 10–49 | | 50– | | Total | |
	First round (%)	Second round (%)	First round (%)	Second round (%)	First round (%)	Second round (%)	First round (%)	Second round (%)
Exhaustion symptoms (3–10)	12	13	9	14	10	15	10	15
Job satisfaction, good or very good	76	79	73	70	73	69	73	70
Poor work capacity, less than 7	6	11	7	12	6	11	7	11
Sickness absenteeism during 12 months, 10 days or more	14	17	14	20	21	21	19	20

was good and informing and communication were well handled, and people felt that they were appreciated. The above organizational practices and climate characteristics were seen here to illustrate the core of a healthy work organization.

In the whole study group the relations between various well-being measures and job and organizational characteristics were also analyzed (Table 6.5). Exhaustion symptoms related to nearly all job and organizational characteristics. The risks in the physical work environment were also slightly related to them. Job satisfaction was mainly related to high job control, appreciation of one's work, and to other organizational practices and climate factors. A high number of risk factors in the work environment, high physical workload, low job control, and low appreciation, showed the strongest correlations to poor work capacity and to sickness absenteeism days, which also correlated negatively with continuous improvement practices (Table 6.5).

Also some personal background and lifestyle factors were correlated to the job and organizational characteristics and well-being. Higher age related statistically significantly to poor work capacity ($r = -0.25$, $p < 0.001$) and lower educational level ($r = -0.27$, $p < 0.001$). Men had more risks in their work environment, and their physical workload was higher than that of women. Self-reported alcohol problems were associated with more exhaustion symptoms and lower work capacity. Frequent physical exercise related slightly to less exhaustion symptoms and better work capacity.

Comparison of job characteristics, organizational practices and climate, and well-being during 1996 and 1997–98

Longitudinal analysis of the employees' results

During the two-year follow-up many changes occurred in the level of perceived job and organizational factors (Table 6.3). Four or more risk factors in the work environment were reported by 29 percent at the first round, and 20 percent at the second. Also the physical workload was reported to be somewhat lower at the second round. Continuous time pressure had increased from 23 percent to 31 percent. A notable decrease was seen in the insecurity concerning job future, which was 28 percent at the first round and 16 percent at the second. No essential changes were found in other organizational practices and climate factors at the level of the whole group.

Sectorially the changes between the branches during the follow-up were also somewhat different. Time pressure increased in all sectors, but most clearly in motor vehicle trade and repair, in hotels and restaurants, and in transport. Supervisory support remained at about the same level in various branches of industry, and the same was true also for co-worker relations,

Table 6.5 Correlation coefficients (Pearson's r) between work and organizational characteristics, and well-being in the first and second study rounds (n = 4068)

	Exhaustion symptoms		Job satisfaction		Work capacity		Sickness absenteeism, days	
	First round r	Second round r	First round r	Second round r	First round r	Second round r	First round r	Second round r
Physical work environment								
risk factors in work environment	0.14	0.17	−0.15	−0.16	−0.17	−0.17	0.10	0.11
Job characteristics								
physical workload	0.07	0.10	−0.22	−0.17	−0.18	−0.17	0.18	0.20
time pressure	0.22	0.24	−0.13	−0.14	−0.10	−0.09	0.01	0.01
job control	−0.22	−0.23	0.33	0.33	0.15	0.20	−0.10	−0.15
Organizational practices and climate								
supervisory support	−0.20	−0.20	0.33	0.34	0.13	0.12	−0.10	−0.07
co-worker relations	−0.21	−0.21	0.24	0.25	0.11	0.13	−0.08	−0.05
continuous improvement practices	−0.17	−0.25	0.35	0.37	0.13	0.16	−0.11	−0.14
information about changes	−0.20	−0.24	0.35	0.35	0.15	0.15	−0.10	−0.12
job future insecurity	0.11	0.10	−0.15	−0.16	−0.03	−0.04	0.00	0.00
positive social climate	−0.21	−0.23	0.30	0.31	0.12	0.14	−0.09	−0.07
appreciation of one's work	−0.23	−0.27	0.38	0.38	0.17	0.17	−0.11	−0.12

Statistical significance:
r = 0.04, p < 0.05; r = 0.05, p < 0.01; r = 0.06, p < 0.001

except for the improvement in hotels and restaurants. There was some decrease in the appreciation of one's work in transport. The biggest change during the follow-up occurred in the insecurity concerning job future; the branches in which the situation had improved most were service stations, metal and engineering industry, construction and transport. These changes indicated the general trend in Finnish worklife during these two measurements. The economic situation improved clearly, and the overall unemployment decreased in the country.

The changes in well-being were also investigated at both study rounds (Table 6.4). A definite increase in exhaustion symptoms was evident: 10 percent had elevated exhaustion symptoms at the first round, but 15 percent at the second round. Also the perceived work capacity of the employees was somewhat lower at the second study round, e.g. the proportion of those experiencing their work capacity as very good decreased during the follow-up from 55 percent to 41 percent. Sickness absenteeism days and general job satisfaction remained at about the same level in the whole group. The changes in well-being during the follow-up were not so clearly related to the size of the enterprises nor to the branch.

Employers' perception of change during the follow-up

During 1996–97 the amount of produced products or services had decreased in 13 percent of the enterprises, but had increased by 10 percent or more in 38 percent of the enterprises. In the same period, the profitability of the enterprise had fallen in 19 percent of the enterprises, but had increased by 10 percent or more in 38 percent of the enterprises. Production had increased especially in construction, manufacturing of electronic appliances, and hotels and restaurants. Profitability had also increased more often in the manufacturing of electronic appliances, motor vehicle trade and repair, and construction. In some branches the number of companies was so small that nothing could be said about the changes. The employers evaluated the work capacity of their personnel to have stayed at about the same level, being good or very good in 75 percent. In 1997 62 percent of the employers noted elevated exhaustion symptoms among the employees at company level, and 51 percent some problems associated with work organization. These problems were not necessarily severe.

When the changes in job and organizational characteristics during the follow-up were summarized, the physical work environment had improved somewhat and the insecurity related to job future had decreased, whereas at the same time, time pressure had increased. However, the organizational practices and climate had remained at about the same level. The employees also reported exhaustion to have increased as well as work capacity to have diminished. These findings among the employees received support from the evaluations of the employers. The findings from the employee and employer follow-up can be concluded to reflect, at the group level, the effects of the simultaneous general increase in economic activity in the whole country, as well as the improved employment situation.

Effects of the planned interventions on job and organizational characteristics and well-being

In order to get a picture of the relations between planned interventions and changes in job and organizational characteristics and well-being, the reported data from both the employers and employees were used.

First the information on the planned interventions based on the employers' reporting was analyzed. When the changes in the well-being of those having more intensive planned interventions were compared to those having only minor or no interventions, the increase in sickness absenteeism days was smaller ($p < 0.05$) among those having more intensive interventions in customer service and collaboration. Also job satisfaction did not decrease, as compared to the others ($p < 0.05$), if the company had carried out interventions in collaboration.

The positive changes in the insecurity associated with job future were related to interventions focused on improving collaboration ($p < 0.05$). However, leadership interventions at the workplace had decreased job insecurity less than at other workplaces ($p < 0.05$). Perceived work climate improved ($p < 0.001$) among those receiving training in multi-skilling, leadership practices, and collaboration. Improvements in co-worker relations were clearly related to the interventions in collaboration ($p < 0.05$). Continuous improvement practices improved slightly among those receiving training in customer services ($p < 0.06$). The participatory approach applied in the intervention correlated statistically significantly with many changes, and most of all with an increase in continuous improvement practices ($r = 0.36$, $p < 0.001$).

The coverage of the interventions carried out was used as the measure when analyzing the relations of the reporting of interventions by the employees and the changes in job and organizational practices (Table 6.6). Although the correlations were weak, some statistically significant relations were found because of the large group size. All types of planned interventions were connected to the improvements in continuous improvement practices ($r = 0.23–0.26$, $p < 0.001$), indicating that the employees perceived that the developmental practices had been more generally adopted than merely as single actions; the climate was thus seen to have become more positive to changes. The organizational climate variable relating significantly to organizational interventions was an increase in the appreciation of one's own work. Also the improvements in informing about changes within the company was slightly related to most of the interventions in question. In addition to these general effects of all interventions,

Table 6.6 Correlations (Pearson's *r*) of the changes in job and organizational characteristics and well-being to the planned organizational interventions

	Planned organizational interventions			
	Customer service	Multi-skilling	Leadership	Colla-boration
	r	*r*	*r*	*r*
Well-being				
exhaustion symptoms	−0.03	−0.04	−0.02	−0.06
job satisfaction	0.03	0.05	0.03	0.05
work capacity	−0.02	−0.06	−0.01	−0.04
sickness absenteeism	−0.00	0.00	0.00	0.00
Physical work environment				
risk factors in work environment	−0.04	−0.02	−0.03	−0.01
Job characteristics				
physical work load	0.01	0.02	0.04	0.03
time pressure	−0.03	0.01	−0.03	−0.03
job control	0.03	0.02	0.04	0.04
Organizational practices and climate				
supervisory support	0.03	0.04	0.02	0.04
co-worker relations	0.04	0.04	0.03	0.06
continuous improvement practices	0.25	0.25	0.23	0.26
information about changes	0.05	0.06	0.05	0.07
job future insecurity	−0.04	−0.01	−0.02	−0.02
positive social climate	0.02	0.04	0.01	0.06
appreciation of one's work	0.05	0.07	0.07	0.10

Statistical significance: $r = 0.04$, $p < 0.05$; $r = 0.05$, $p < 0.01$; $r = 0.06$, $p < 0.001$

the social climate and co-worker relations had benefited from interventions
dealing directly with the collaboration within the company. Changes in the
physical work environment or in physical workload were not related to any
type of intervention, nor to time pressure or control.

The relations between the changes in well-being and employee-reported
interventions were very few. Work capacity had improved somewhat positively along with the multi-skilling of workers ($r = 0.06$, $p < 0.01$), and
exhaustion symptoms decreased when the interventions were focused at
collaboration ($r = 0.06$, $p < 0.001$).

The effects of the interventions, based on the employers' reporting,
were most pronounced on the improvement of the social climate, and
somewhat less on the continuous improvement practices and sickness
absenteeism days. The interventions as reported by the employees themselves were all related to an increase in continuous improvement practices,
in informing workers about changes, and in the appreciation of one's work.
Changes in work capacity and exhaustion symptoms were only slightly
related to the intervention carried out.

Explaining the changes in well-being and other changes during the follow-up

The changes in job and organizational practices correlated slightly to the changes in well-being in the whole group (Table 6.7), although the direct effects of an organizational intervention on well-being cannot be expected to be very notable. The decrease in exhaustion symptoms correlated slightly positively to improvements in nearly all job and organizational characteristics, and most of all to appreciation, continuous improvement practices and other climate factors, but also to time pressure and job control. The increase in job satisfaction related to positive changes in appreciation, supervisory support, social climate and informing about changes. Improved work capacity was related to increased job control, and lower physical workload. The change in sickness absenteeism was related very slightly to these changes in work and organizational characteristics. Also the changes in lifestyle were examined in relation to changes in well-being. The decrease in self-reported alcohol problems related slightly to better work capacity, and increased physical exercise related slightly to less exhaustion symptoms and sickness absenteeism days.

It can be concluded that improvement in organizational practices and climate affected positively exhaustion and job satisfaction, while improvement in job characteristics and physical workload affected positively work

Table 6.7   Correlations between changes during the follow-up in well-being, and job and organizational characteristics (Pearson's correlations) (*n* = 4032)

	Exhaustion symptoms r	Job satisfaction r	Work capacity r	Sickness absenteeism days r
Physical work environment				
risk factors in work environment	0.09	−0.09	−0.09	0.02
Job characteristics				
physical work load	0.09	−0.09	−0.17	0.07
time pressure	0.14	−0.07	−0.10	0.02
job control	−0.13	0.15	0.16	−0.02
Organizational practices and climate				
supervisory support	−0.15	0.16	0.11	−0.02
co-worker relations	−0.10	0.12	0.07	0.00
continuous improvement practices	−0.15	0.12	0.12	−0.03
information about changes	−0.13	0.17	0.10	−0.03
job insecurity	0.08	−0.09	−0.06	0.02
positive social climate	−0.14	0.16	0.11	−0.03
appreciation of one's work	−0.16	0.18	0.13	−0.04

Statistical significance: *r* = 0.04, *p* < 0.05; *r* = 0.05, *p* < 0.01; *r* = 0.06, *p* < 0.001

capacity. Improvement in lifestyle factors affected positively, though slightly, health-related measures of well-being.

Changes in the two main measures of well-being were explained by using stepwise multiple regression analysis. Eight percent of the variance in the increase in exhaustion symptoms could be explained by an increase in physical risk factors, continuous time pressure and a decrease in continuous improvement practices and appreciation of one's work, as well as by a lack of the intervention promoting collaboration ($R^2 = 0.08$, $F = 30.79$, $p < 0.0001$). Twelve percent of the increase in job satisfaction was explained by increased job control and continuous improvement practices, better informing about changes, and better co-worker relations, as well as by interventions directed at supervisory practices ($R^2 = 0.12$, $F = 45.13$, $p < 0.0001$).

Relative company effectiveness as compared to employee well-being, job and organizational characteristics and the interventions carried out

The relative effectiveness of the companies which reported at least some organizational or training interventions was compared to the relative effectiveness in those with no or only minor interventions. The higher relative productivity in 1997 was related to the interventions in the multi-skilling of personnel and in leadership practices. The higher relative profitability in 1996 and 1997 was related to the more intensive interventions of collaboration within the company, and in 1997 also to the more intensive intervention of leadership practices and multi-skilling of workers. These same relations to company effectiveness were also examined correlatively (Table 6.8). Good profitability in 1997 was related statistically significantly to all kinds of organizational development interventions at workplace. But the high production in 1997 was related only to interventions dealing with the multi-skilling of workers and the development of managerial practices. The lower age of the company was related positively to both of these effectiveness measures, while the number of employees in the enterprise was related positively to the amount of production.

The employees' perception of their job and organizational characteristics and well-being in 1997 was associated with company effectiveness. The employees' job satisfaction was the only measure of well-being that was positively related to the level of production and profitability of the company. Among job and organizational characteristics, good supervisory support and good social climate were related statistically significantly to the level of production and profitability. Also continuous improvement practices related to the level of production. These organizational factors were the ones which had also been promoted successfully by the aforementioned planned interventions.

Table 6.8   Correlations (Pearson's r) of perceived relative productivity and profitability to job and organizational characteristics, well-being and planned interventions at company level ($n = 218$)

	Productivity 1997 r	Profitability 1997 r
Age of enterprises	−0.16	−0.18
Number of employees	0.18	0.10
Well-being		
exhaustion symptoms	−0.09	−0.11
job satisfaction	0.17	0.21
work capacity	0.10	0.07
sickness absenteeism	0.01	−0.06
Physical work environment		
risk factors in work environment	0.00	0.04
Job characteristics		
physical work load	0.09	0.08
time pressure	0.11	0.06
job control	0.08	0.11
Organizational practices and climate		
supervisory support	0.17	0.24
co-worker relations	0.05	0.02
continuous improvement practices	0.19	0.19
information about change	0.07	0.11
job future insecurity	−0.10	−0.12
positive social climate	0.23	0.18
appreciation of one's work	0.11	0.12
Planned organizational interventions		
customer service	0.11	0.18
multi-skilling	0.19	0.21
leadership practices	0.19	0.18
collaboration	0.13	0.24

Statistical significance: $r = 0.04$, $p < 0.05$; $r = 0.05$, $p < 0.01$; $r = 0.06$, $p < 0.001$

In stepwise multiple regression analysis, 13 percent of the variance of the relative productivity in 1997 was explained by good social climate at work, higher number of workers, and young age of the company ($R^2 = 0.13$, $F = 7.11$, $p < 0.0002$). Twelve percent of the relative profitability was explained by young age of the company, higher number of employees, and good supervisory support ($R^2 = 0.12$, $F = 6.91$, $p < 0.0002$). The planned interventions carried out were not used as explanatory variables because they can also be viewed as effects of improved productivity and profitability.

DISCUSSION

This study described how organizational practice and climate factors which have been linked to the healthy organization model were related to workers' well-being in SMEs. The main aim was, however, to analyze the changes that took place in job and organizational characteristics and well-being during a two-year follow-up, and how these changes were related to the planned organizational interventions. Special attention was directed toward the relations between company effectiveness and workers' well-being and perception of their job and the organizational characteristics.

Small company size, as defined by the number of employees, was a central determinant of good organizational practices and climate. The well-being of workers did not vary according to company size; only employees from enterprises with more than 50 employees had more sickness absenteeism days. The differences between branches of industry were clear in many job and organizational characteristics. Each of the measures of well-being related somewhat differently to job and organizational characteristics. For example, exhaustion symptoms were related to all of them. This differed somewhat from what has been found with the same methods in bigger companies where time pressure has had the main impact on exhaustion symptoms (Lindström 1997). Job satisfaction was related to organizational practices and climate, whereas perceived poor work capacity and high sickness absenteeism were more dependent on risks in the physical work environment, higher physical workload, low job control and low appreciation of one's work.

The organizational interventions planned at the companies were based on their own expressed needs and on the feedback they got from the results of the first survey. The interventions were carried out by the companies with consultative support from the project group, OHS personnel or some other consultants.

The changes noted in the whole study group between the two surveys clearly reflected the general trends in Finnish worklife during the follow-up period. The perceived future insecurity of the job decreased clearly as the employment situation improved in the whole country. Continuous time pressure and exhaustion symptoms increased also as in Finnish worklife, reflecting the simultaneous general trends of reorganizing business processes and intensifying the work.

In order to learn more about the effects of the various kinds of organizational interventions, their effects were analyzed. The perception of the coverage of interventions differed between employers and employees, the view of the employers being more positive. The various interest groups do not always view changes at work from their own perspective (Birchall et al. 1978). Interventions reported by employers in customer service, leadership practice and collaboration showed positive effects on

the employees' perception of the general work climate and co-worker relations. Future insecurity of the job was slightly positively affected by
collaboration interventions, but negatively by leadership practice interventions. Sickness absenteeism days decreased slightly when improvements took place in customer services and collaboration, i.e. in social
interaction in general.

The interventions as reported by employees produce somewhat different
results. Continuous improvement practices were found to be mostly
improved by all kind of interventions. This is quite natural and indicates
that the interventions carried out have been well accepted, and especially
the continuous improvement practices have been rooted in the organization (Starkey 1998). Also the changes in appreciation and better informing
about changes were related to the interventions. The collaboration
promoting interventions had improved the social climate and co-worker
relations. The changes in well-being which associated directly with the
interventions were very slight. The changes in job and organizational
characteristics and in well-being during the follow-up were clearly interdependent. In particular, changes in exhaustion symptoms and job
satisfaction were related positively to improvements in most job and organizational characteristics. Work capacity benefited especially from
improvements in physical work factors and job control. Similar effects
have been noted in an earlier follow-up study among municipal employees
(Tuomi *et al.* 1997).

Both the reported relative productivity and profitability of the company
associated positively with the interventions carried out. One can say that,
in general, companies with greater resources are more likely to carry out
various kinds of improvements, and also to employ successful strategies
to take care of the future and their human resources, i.e. the workers. In
addition, the job satisfaction of workers was positively related to higher
productivity and profitability. Good supervisory support and positive
work climate were also factors contributing to company effectiveness, and
the continuous improvement practices contributed to the higher level
of production (Cooper and Cartwright 1994). In explanatory analysis
the young age of company and higher number of employees were both
relative effectiveness measures. In addition, the good work climate
explained productivity and the good supervisory support the profitability.
The variances were low, but statistically significant.

Evaluation of the intervention and survey methods

The content and process of interventions carried out were not described
in detail; only the type and intensity were assessed by the employees and
employers. This is lacking in the present study, but this analysis can be

seen as an exploratory phase, which should be deepened. The results showed, however, that the interventions as reported by the employees were more focused, and related to well-being and organizational practices and climate. The results of the follow-up in general are valid because the changes that took place in this same period in Finnish worklife were also evident in the present study group. The positive issue concerning the interventions was the simultaneous general trend towards a better economic situation in the whole country after the recession in the early 1990s. This created a positive atmosphere for the development of organizations.

The interventions being evaluated were organizational ones, as the individual-directed interventions could not be traced when dealing with so many companies and employees. However, these were also carried out with the help of OHS personnel, but the indicators and design used here were too crude to analyze them.

The selected well-being measures covered different aspects of well-being and formed a fruitful set of measures. On the job and organizational side, however, a better measure of work pressure would have been required. It was somewhat surprising that continuous time pressure, so common in Finnish worklife today, did not display a stronger relation to well-being or to the interventions. Perhaps the objective workload and weekly working hours would have been better measures of quantitative workload. Our emphasis, however, was on organization practices and climate measures because the planned interventions were focused on these aspects. Interviews of employers and employees might have increased the validity of the changes resulting from the interventions.

The healthy work organization model revisited

If we start from the assumption that a healthy work organization combines the well-being of the employees and company effectiveness, the results are promising. Company effectiveness was explained by supervisory support and a positive social climate; also job satisfaction contributed to company effectiveness. In interventions, this encourages one to improve supervisory practices and to promote the work climate, and also to see them as factors influencing company success. The continuous improvement practices were seen to improve via all kinds of planned interventions and to be related to good productivity at company level. Appreciation of people has often been seen as the basis for a healthy organization (Jaffe 1995). Also in this case it was associated with the success of the planned interventions and the well-being of the employees. The appreciation is probably more a consequence of good organizational practices, such as supervisory support, informing about changes, and continuous improvement, in the same way as a good social climate. In conclusion, these interventions increasing the competence of workers (multi-skilling and customer service), and the development of leadership practices and collaboration had positive effects; first, on organizational practices, and through them on the well-being, and company effectiveness associated with good supervisory support and good organizational climate. The link between company effectiveness and the interventions carried out can be two-fold. The successful companies invest in workers and the improvement of the organization, which, in turn, affect profitability. Small company size was thus beneficial, as regards the healthy company ideal, but effectiveness was better in the medium-sized companies.

References

Birchall, D. W., Carnall, C. A. and Wild, R. (1978) The development of group working in biscuit manufacture – A case study, *Personnel Review* 7: 40–9.

Cooper, C. L. and Cartwright, S. (1994) Healthy mind; healthy organization – a proactive approach to occupational stress, *Human Relations* 47: 455–71.

Cox, T. and Leiter, M. (1992) The health of health organizations, *Work and Stress* 6: 219–27.

Huuskonen, M. S., Koskinen, K., Bergström, M., Vuorio, R., Järvisalo, J., Ahonen, G., Forss, S., Järvikoski, A., Lindström, K., Roto, P., Ylikoski, M. and Rantanen, J. (1997) Working capacity as a success factor in a small enterprise and in society, *Work and People, Research Report* 10: 209–20.

Jaffe, D. T. (1995) The healthy company: Research paradigms for personal and organizational health, in S. L. Sauter and L. R. Murphy (eds) *Organizational Risk Factors for Job Stress*, Washington, DC: American Psychological Association.

Levering, R. (1988) *A Great Place to Work*, New York: Random House.

Lim, S. Y. and Murphy, L. R. (1997) Models of healthy work organization, in Proceedings of the 13th Triennial Congress of the International Ergonomics Association, Tampere, Finland. *From Experience to Innovation*, pp. 501–3, Helsinki: Finnish Institute of Occupational Health.

Lindström, K. (1997) Assessing and promoting healthy work organization, in Proceedings of the 13th Triennial Congress of the International Ergonomics Association, Tampere, Finland. *From Experience to Innovation*, pp. 504–6, Helsinki: Finnish Institute of Occupational Health.

Starkey, K. (1998) What can we learn from the learning organization? *Human Relations* 51: 531–46.

Tuomi, K., Ilmarinen, J., Martikainen, R., Aalto, L. and Klockars, M. (1997) Aging, work, lifestyle and work ability among Finnish municipal workers in 1981–1992, *Scandinavian Journal of Work, Environment and Health* 23: suppl 1, 58–65.

Part II

Company-wide policies and programs

The chapters in this section describe broad-brush, company-wide programs and policies designed to improve both employee health and organizational effectiveness. Adkins, Quick, and Moe describe a US Air Force effort that began with a set of four organizational health principles modeled after definitions of individual health. These principles were:

1   Health is more than the absence of disease.
2   Health is a process, not a state.
3   Health is systemic.
4   Health requires positive collaborative relationships.

Based on these principles, a model of organizational health was proposed which guided the development of new programs and assessment tools to promote worker health and organizational effectiveness. Schmidt, Welch, and Wilson describe a broad and varied set of programs and services within Home Depot called *Building Better Health*. Elements of *Building Better Health* include traditional lifestyle/health promotion programs such as smoking cessation, diet/nutrition, health screening, and stress management, but also services dealing with healthcare benefit plans, child and elder care, family leave, and community volunteering activities. The *Bringing Health to Life* program at Zeneca (now AstraZeneca) pharmaceuticals in Great Britain recognizes the challenging and competitive nature of modern work and seeks to lower employee stress and improve organizational performance. Teasdale, Heron, and Tomeson present an organizational success/individual well-being model that attends to both worker needs and organizational functioning. The final chapter in this section describes a "work in progress" at Britain's largest retailer, Marks & Spencer Plc. The chapter offers an inside look at how a *Managing Pressure* program is conceptualized and implemented within a large company. Although this work is in its early stages, and evaluative results are not available, the chapter should be especially useful to readers who are initiating new programs because the authors set forth in great detail the necessary steps in the process of conceptualizing the program, identifying the need, establishing a project team, selecting an intervention model, and strategically positioning the program within the organization.

Chapter 7

Building world-class performance in changing times[1]

J. A. Adkins, J. C. Quick, and K. O. Moe

The United States Department of Defense (DOD) shares the quest for vibrant organizational health with other public and private organizations around the world. A successful quest requires managing potential threats, most significantly those associated with occupational stress. An emerging new organizational reality is transforming the industrio-political landscape for the twenty-first century and is increasing occupational stress for individuals, groups, and organizations at work, with an accompanying sense of threat (Gowing *et al.* 1998; Staw *et al.* 1981). Stress is a central issue in any dramatically changing environment, be the change significant growth and expansion, as in the case of ICI-Zeneca in the pharmaceutical industry (Teasdale and McKeown 1994), organizational disaster, as in the case of Union Carbide at Bhopal (Shrivastava 1987), or organizational retrenchment, as we discuss in this chapter. In each of these three scenarios, organizations must ensure systems are in place to guard the health and vitality of the organization and its members and to successfully navigate through the changes (Adkins 1996). Organizations can reasonably expect excellence despite the threat of change and stress (Steely 1999). Our thesis is that opportunities do exist and others can be created to build world-class performance even in the midst of organizational turbulence.

In no environment have change and transformation been more prominent and rapid than in the world's armed forces. The dramatic military force reductions in the United States and other countries bring to the forefront concerns about stress and threat related to individual and organizational health. Over the last decade, the United States Air Force, as part of the DOD, has undergone an unprecedented level of reorganization (Cliatt and Stanley 1994). The end of the Cold War foretold a number of changes in military force structures around the world. The changing geopolitical environment has been accompanied by a reduction in the number of military personnel. Between 1991 and 1997, the total number of active duty US Air Force members declined by 36 percent. The British Royal Air Force (RAF) declined by about 40 percent during the same period, with a parallel decrease in Canadian Forces. Likewise, Russia is in the process of reducing its military force structure from 1.5 percent of its population to 1.0 percent. Associated with personnel reductions, the US Air Force and the RAF also reduced infrastructure, including closing and consolidating bases in connection with shrinking resources (Adkins 1998).

Within the US Air Force, the overall structure of the organization also changed, bringing with it changing terminology and subgroup identification. Although the number of personnel and resources declined, the demands on time and resources actually increased in keeping with a changing mission associated with escalating worldwide regional conflicts and humanitarian tasks. With the closure of permanent overseas bases, requirements for temporary deployments, or work away from home base, increased four-fold overall, with some highly specialized units seeing an increase in deployments five times that prior to the drawdown efforts. Individual members remaining at home base were tasked with additional duties left behind, often picking up responsibilities not traditionally considered a part of their primary job responsibilities or training, leading to an increase in role ambiguity. Many of these responsibilities were administrative or non-technical in nature, which promoted an increase in perceived skill under-utilization combined with high levels of work overload. Privatization, outsourcing, and dual use initiatives brought civilian employees and contractors into the workplace along with changing rules associated with job completion and resource allocation. Benefits, including healthcare, retirement benefits, and training opportunities, declined as resources were truncated. Cultural identity with established military traditions was threatened by the changing organizational structure, increased civilianization, changes in the original "contract" between the members and the American public, and increased integration with other military services and Nato peacekeeping forces. These multiple changes persisted, overlapping in their time and effect. With such change comes stress.

The concerns of defense organizations in response to stress are similar to industrial organizations engaged in corporate warfare; namely, that stress has the potential to negatively impact people, organizational effectiveness, or military readiness, and safety (Nelson *et al.* 1989). To build a healthy, vital organization despite turbulence in the system, military personnel integrated concepts from preventive medicine and public health with practices in clinical, health, and organizational psychology along with management and safety sciences to form a multi-faceted, multi-disciplinary approach designed to identify both potential hazards and protective factors within the organization and to quantify risks associated with occupational stress (Adkins 1999; Last and Wallace 1992; Quick 1999b). This process of risk identification and assessment is a fundamental step toward the ultimate goal of developing countermeasures to mitigate and manage those risks (Ordin 1992). This process has taken place on multiple levels within this large organization, with some initiatives taking place on a global, system-wide basis and others being targeted to smaller individual units.

This chapter highlights initiatives in the US Air Force branch of the DOD to both assess and mitigate psychosocial risks in efforts to build and maintain world-class performance within a healthy organizational context. We begin with a framework of preventive management for organizational health (Quick 1999a), then add an operational risk management model for defining hazards, assessing risks, identifying risk mitigation strategies, and tracking results of intervention strategies. Examples of data used in quantifying risks and tracking change at various levels within the organization are provided throughout.

ORGANIZATIONAL HEALTH PRINCIPLES

For centuries, humankind has struggled with issues of individual health – how to define it, assess it, protect it, repair it, and maintain it. In recent history, organizational health has become an important, related topic. As individual and organizational health are interdependent (Quick *et al.* 1997), setting out some basic principles of individual health can help us define organizational health.

Principle One: Health is more than just the absence of disease

Health runs along a continuum ranging from the fullness of life, or vibrant well-being, through chronic disease or severe distress to, ultimately, absence of life, or mortality. Thus, for organizations the goal is to move toward world-class performance and abundant organizational health, not just to reduce distress or organizational dysfunction or to avoid organizational demise. Promoting organizational health, therefore, requires identifying various stages and indicators of health along the continuum.

Principle Two: Health is a process, not a state

Rather than a static state of being, health is a complex, dynamic process. To "have" good health requires continual attention and effort. Once a level of optimal health is achieved, the process of health maintenance must be continued or health atrophies. Because the process of health is continually changing, vigilant awareness, or surveillance, and recurrent assessment are necessary to ensure that the process continues on-track. Surveillance is needed on multiple levels to match process complexity. Surveillance information is transformed into awareness through various risk communication processes. Without awareness on the part of process owners, surveillance and assessment data, no matter how accurate, mean little to the health process.

Principle Three: Health is systemic

Health involves interrelated parts working together in balance (Cannon 1932, 1935). If one part becomes diseased or dysfunctional, the entire system suffers in terms of decreased potential functioning capability, increased workload to compensate for the diseased part, potential contamination, and increased vulnerability. Individuals vary in their level of vulnerability to disease, as do organizations. Building immunity to external threats requires identification of vulnerabilities along with available resources within the system for fending off dysfunctional processes, and then building behavioral inoculation programs around those identified resources and deficiencies. The level of risk to the organization, and to individuals within the organization, may be quantified through comparing the threats with the vulnerabilities. High-risk behaviors and stress-inducing psychosocial factors within the organization can be identified and modified to mitigate potential worklife problems and to optimize systemic well-being.

Principle Four: Health requires positive collaborative relationships

The burden of suffering on the public health battlefield has shifted from infectious and communicable diseases early in the century to chronic disorders in which behavioral and lifestyle factors play a central role (Cohen 1985; Foss and Rothenberg 1987). Gone are the days when the medical profession was revered for superhuman ability to heal all disease. Multidisciplinary teams of healthcare providers increasingly are assuming roles as trained and knowledgeable consultants to patients who participate in decision-making regarding their health. The same principle applies to the organizational health environment. In an increasingly information and service-oriented work environment, organizations thrive when they find ways to develop and nurture positive relationships with employees, customers, and suppliers. The nature of those relationships may change, as organizational cultures evolve through downsizing, outsourcing, and changing employment contracts, but the need to build and maintain the relationships through changes remains important to the well-being of all interrelated parties. Building human capital has become an important part of organizational management that requires leaders to identify individual and team strengths and limitations and to design operations based on those resources. Leaders and managers cannot fix the organization any more than a healthcare provider can fix a patient's high-risk behavior. They can only catalyze action through relationships and communication. Likewise, organizational health practitioners serve in a collaborative role with organizational members – executives and workforce – and they network with, or consult, other professionals to augment their knowledge and skill.

ORGANIZATIONAL HEALTH PROCESS

In consideration of these principles, organizational health may be considered a dynamic, systemic process dependent on relationships both internal and external to operations, as depicted in Figure 7.1. Maintaining health requires assessment and surveillance of potential threats to health, organizational resources, immunity or vulnerability to those threats, and indicators of health. This information is most effective when it comes to the awareness of organizational health process owners. Organizational health promotion, prevention, and intervention strategies depend on information from surveillance and risk communication processes to move the organization toward the ultimate goal of world-class performance, with its hallmark of excellence.

The organizational health management model is most effectively applied using a flexible and adaptive process that can supersede content areas as times and conditions change. This process is depicted as an encircling

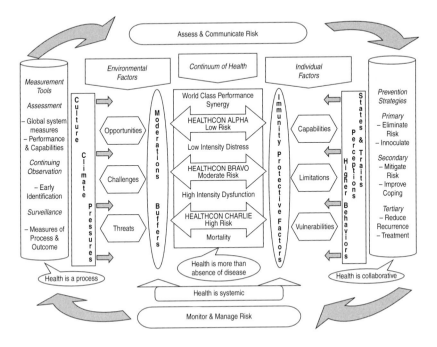

Figure 7.1 Organizational health management model

system in Figure 7.1. It begins with assessment – identifying, defining, and operationalizing concepts and developing appropriate measurement tools. The measures should lend themselves to quantifying frequency, intensity, duration, proximity, and probability of occurrence of high-level performance indicators along with internal and external risks to organizational health. Results of the assessment process are subsequently communicated to process owners, usually organizational leaders, who can examine both assessed risks and available preventive strategies. In consultation with organizational health and assessment personnel, leaders make judgments that match preventive strategies with identified capabilities and risks and implement change procedures and programs, as needed. Both process and outcome measures of these programs are monitored along with continued surveillance of organizational health conditions to ensure the process is on track. In ongoing surveillance, additional risks may be identified which trigger the process to continue.

Within the operational Air Force, this process has been implemented through Operational Risk Management (ORM). ORM was championed by the Air Force Safety Center and implemented as an Air Force-wide program not only to guide identification and mitigation of risk within a rapidly changing operational environment, but also to maximize organizational performance. The process, adapted from a successful US Army strategy, represents a systematic, data-based approach to identifying and controlling hazards and risk, and is equally applicable to both operational and behavioral risk factors. It is important to keep in mind that the mission of the armed forces is inherently one of high risk. Therefore, elimination of all risk is both unlikely and undesirable. Rather, the goal is to find the optimal level of risk matched with mission requirements and available protective factors.

In implementing ORM, all Air Force members received risk management training with the intent of increasing awareness and skill on the part of each individual to identify and mitigate risk in their own personal life and in their work area. Every individual is thus seen as a risk manager. If an individual identifies a risk that is broader than or falls outside his or her scope of responsibility, then the risk is communicated upward in the organization so that the risk management process may begin within the larger system. The process is separate from content and can be applied at all levels for all areas of potential risk. The ultimate goal is to institutionalize this process into the culture of the organization so that all personnel make risk management a way of life.

THE CONTEXT OF HEALTH

Health risks are associated with several elements of the system. Psychosocial risks in the system are generally identified in terms of stress.

Stress is a creatively ambiguous word used with great imprecision. Definitions abound. We assume that occupational stress is a process beginning with a transaction between environmental and individual factors within the work environment and ending with either a positive or negative outcome. Increasingly positive outcomes move us toward higher levels of health. Referring to the model in Figure 7.1, stress measurement and intervention may involve focusing on any or all of these variables. Environmental pressures and their associated risks can be reduced, eliminated or modified while system-based buffers or moderators to pressures can be developed or nurtured; and, a health-engendering environmental context can be developed by establishing a healthy culture and climate (Weick 1987). Individual vulnerabilities can be reduced and individual strengths or protective factors can be enhanced. In addition, the fit between the environment and the individual can be a target for assessment and change. When the environmental factors match well with individual capabilities, a condition of low risk is achieved. As the pressures face limited coping abilities of individuals they present an increasing level of risk or challenge until vulnerable individuals perceive and experience a high level of threat to their functioning and well-being in response to environmental factors. Lastly, potentially negative outcomes can become the focus of measurement and intervention when neither environmental nor individual antecedent factors can be modified either separately or in combination.

Stress is a multi-dimensional process, originally labeled the emergency or fight-or-flight response about the time of World War I (Cannon 1929). The introduction of the concept of a general adaptation syndrome (GAS) extended our understanding of the stress process and focused attention on the distinction between acute and chronic stress responses (Selye 1973, 1976). This distinction is equally applicable at the organizational level. Individuals who are persistently exposed to an unrelenting high stress environment are likely to experience different outcomes than those who experience acute, short-term high stress events. Organizations that create or permit the occurrence of even moderately high levels of stress over long periods of time without respite for its members will experience different outcomes in terms of organizational measures of effectiveness than organizations that experience periodic surges in stress. The characteristics of a healthy organization, then, would tend to differ depending on the nature of the stressors within its organizational context.

The organizational context is the fabric into which the elements of stress are woven. Attempts to promote organizational health or to manage change and stress have less opportunity for success if managers, consultants, or other change agents fail to attend to important aspects of the organizational environment – physical, temporal, and psychosocial. An approach appropriate for one organization, or a subgroup within a larger organization, may fail miserably in another of the same size and workforce composition because of differences in culture, climate, or general atmosphere.

To measure and intervene in stress within the US Air Force, several processes were set into place, varying with different levels of health. These processes were in addition to the routine strategies that serve continuously to manage organizational effectiveness and military readiness. The Air Force is a large and complex organization with many programs at different levels within the system. Therefore, this discussion highlights only newly emerging programs.

LEVELS OF HEALTH

The process of managing organizational health risks entails reducing risk to the lowest possible level to enable the organization to thrive. Levels of risk can be associated with levels of health. In the military system, operational risks are referred to as threats and levels of risk are communicated as threat conditions, or Threatcons. Once the existence and capability of a potential threat is identified, the system is triggered to identify levels of associated risk, signaling the need for increased alertness to the potential threat. Threatcon Alpha signifies the lowest level of identified operational risk, followed by increasing risk associated with Bravo, Charlie, and Delta. Applying that general terminology to the realm of health, we can refer to levels of organizational health as health conditions or Healthcons. Changing times represent clear and present threats to organizational health requiring elevated alertness to potential risk factors. Therefore, levels of risk with their associated countermeasures can be used to anchor our health continuum.

Healthcon Alpha

The highest level of health describes an organization with high levels of performance and correspondingly positive levels of individual well-being. The capabilities of individual personnel are well matched to the pressures and culture of the organization. Thus, pressures are perceived as opportunities and confronted with enthusiasm and high levels of energy. Individuals work with the environment, rather than against it, and the two collaborate to generate a synergistic effect to produce world-class performance. Threats, or potential hazards, are present; nevertheless, risk is low. At this level, health involves much more than just preventing problems. A key element of performance is the way the organization builds, maintains, and extends its relationships on every dimension, which in many ways is a measure of both the culture (or personality), climate (or mental status), and functioning (or behavior) of the organization. Assessment strategies are global and designed to measure the capabilities of individuals and the organization. Primary prevention strategies aim to continuously improve and maintain the level of health through nurturing a health-engendering work environment. Naturally, there is no one right way to achieve world-class performance, even within the same organization. Times and conditions change. Therefore, to sustain high levels of organizational health, the organization, including the surveillance and prevention strategies, must be capable of change as well. Adaptability, or the long-term accommodation to seismic forces, and flexibility, or the short-term capacity to flex to a crisis, are critical components for program development and for monitoring.

Surveillance strategies

Strategies for assessment focus on measures of capabilities, function, and structures. Both active and passive ongoing surveillance measures attempt to monitor performance, military readiness, and operational functioning to maintain or improve high levels of effectiveness. Measures of relationship, commitment, communication and information exchange, and capacity to flex with change are important and can be incorporated into assessments of corporate culture, employee commitment and innovation, and organizational climate.

A variety of routine readiness and performance measures are tracked at every level within the Air Force. Many of these are specific to individual unit missions and objectives. They are rolled up from the work unit to higher levels of middle management and to upper Air Force leadership for monitoring and planning. Data are electronically stored and are available in near real time through improved information technology. In addition, Functional Management Reviews cut across different organizations to look at processes that are common to multiple work units. Air Force-level personnel survey, interview, conduct task analyses and time measures, and collect a variety of other data points at all locations engaged in the same process. This review compiles best practices to be shared across unit boundaries to ensure the organizational process is the most effective and efficient it can be.

Organizational Health Assessments (OHAs)

OHAs were designed specifically to provide a comprehensive process to assess organizational health (Adkins 1999). The methodology consists of collecting both quantitative and qualitative data specific to the organization. These data sources are specifically designed to measure culture, climate, and functioning and are conducted with individual units within the Air Force system. The OHA process begins with a consultation with a unit's primary leadership and support personnel to identify specific questions for examination and to gain a better understanding of the vision, mission, and atmosphere of the organization. Consultations with unit support personnel also aid in providing information about the resources available to the organization to promote adaptability to change. Formal data collection begins with written questionnaires that include both quantitative and qualitative measures. A team of trained personnel then conducts semi-structured interviews and focus groups with unit personnel to collect organizationally based information. All responses, written and oral, are collected anonymously. These data are tabulated and analyzed before being reported to the process owner – the unit commander who requested the assessment. The team then assists the commander in preparing an action plan and a briefing of the final results to present as feedback to the unit. The OHA provides a comprehensive, individualized description of the capabilities and limitations of the unit along with behavioral recommendations for enhancement of strengths or mitigation of risk. Follow-up assessments are conducted as often as needed to assist with implementing the plan and monitoring change processes.

Primary prevention strategies

Routine strategies

Identification of mission, vision, core values, and competencies lies at the heart of effective organizational health initiatives. People need to know what they are doing and why. Reducing ambiguity through strategic planning and communication of these fundamental aspects of the organization form the foundation of organizational effectiveness.

The best way to ensure high capabilities of personnel is at the point of selection, an element of organizational health often taken for granted. Selection relies on predetermined core capabilities and behavioral attributes that represent the highest probability of success in a career field. The Enhanced Flight Screening (see King, in press; King and Flynn 1995) program was designed to improve the selection processes for pilots. Training pilots is an expensive endeavor. Selection of those individuals with the highest potential for success decreases training costs and maximizes the availability of new pilots emerging from the training pipeline. Similarly, the Behavioral Assessment Service (Cigrang et al., in press) works with new recruits at basic training to identify individuals who are a poor match for military service. Similar programs are available for high stress jobs, such as military training instructors and recruiters as well as some special operations jobs, and for highly sensitive jobs involving classified data. Selection in personnel drawdown programs is conducted through force-shaping. Selected career fields are targeted for reductions while ensuring that critical core competencies and experience levels are maintained.

The Organizational Health Center (OHC)

The OHC was founded at the Sacramento Air Logistics Center in McClellan Air Force Base, CA (Adkins 1999). The specific mission of the OHC was to maximize health – physical, psychological, behavioral, and organizational – by applying behavioral science technology to a workplace setting. This systemic organizational health program employed six program components to implement the four strategies for reducing psychological disorders at work (Sauter *et al.* 1990) as proposed by stress researchers at the National Institute for Occupational Safety and Health (NIOSH). Organizational change was addressed through coordination of community services and consultation on various other aspects of the organization, including change management, teambuilding, and management coaching. Information, education, and communication for workers were enhanced through information broker services and targeted training and prevention programs, such as workplace violence prevention and suicide risk reduction. Enriched psychological services were provided through a variety of work site-oriented support programs, including employee assistance coordination, work site wellness, and peer counseling. Surveillance and monitoring were accomplished through collection and reporting of indicator data.

The OHC Director was assigned a role similar to that of a chief psychological officer, reporting directly to the chief executive officer. Nevertheless, the Director functioned as an internal consultant with the responsibility of looking after the healthcare of the organization as a whole rather than being identified with any single group – management or labor – within the organization.

Various process and outcome measures were used to evaluate effectiveness and success in meeting program objectives. At the end of the first year, workers' compensation rates declined by 3.9 percent (following a 4.6 percent increase the previous year), exceeding the management-established goal of a 3 percent reduction, saving over $289,000 in workers' compensation costs. Healthcare utilization rates declined by 12 percent, yielding a saving of over $150,000 in recaptured productivity alone. Deaths resulting from behavioral problems, including suicides, declined by 41 percent, resulting in cost savings of over $4 million. Critiques, solicited for all services, reported excellent ratings of satisfaction. Because of success in the first year of operations, the program was both continued and expanded by the organization's corporate board despite shrinking resources during federal downsizing and base closure.

In addition, OHC programs were established at other Air Force bases and some individual program components were replicated for the entire Air Force population.

Readiness Assessment and Planning Tool Research (RAPTR)

The RAPTR project (Kuper and Barlow 1998) was designed to develop and demonstrate advanced technology to assist Air Force logistics agencies in the preparation, planning, and managing of organizational changes and process improvements.

Over 70 percent of all process change efforts fail (Wellins and Murphy 1995). Important organizational issues such as cultural, technological, and strategic issues within the organization are often ignored and tools and methods for planning and implementing change are often fragmented. As a result, although organizational change occurs with increasing frequency, both organizations and individuals commonly fail to benefit from lessons learned in their experiences with change. RAPTR stepped in to remedy that situation through correcting the deficiencies in available change management tools.

The RAPTR methodology sought to provide a user-friendly tool for managers to assess their organization's culture, technology, and readiness for change and to develop a detailed plan for process change. Data were captured in real time through computerized assessment instruments and transferred into a modeling and simulation program. The simulation tool provided a knowledge base to guide change agents in both planning and overcoming resistance to change. Information from the assessment tool, including goals and objectives, organizational culture, work-unit specific issues and technology, was funneled into a Change Designer's Notebook. The computerized simulation integrated the information and simulated potential outcomes based on prior research taken from the literature and previous lessons learned and entered into the simulation model. The Notebook then continued to guide the manager through the change process as parameters were modified. The end result was an integrated process re-engineering tool that brought all the necessary information into one place for easy use by managers in planning and implementing change.

Healthcon Bravo

The second level of health represents the average day-to-day functioning for many organizations. It corresponds with a condition that is absent disease; organizational functioning falls within routine levels of effectiveness. Typical operations, even in an overall healthy organization, often lie within this level of the health continuum. At this level, increasing pressures connect with limitations in coping and are perceived as challenges or barriers to effectiveness rather than opportunities to excel. As these challenges are confronted, some low-level distress may appear in the system. The level of risk is moderate. Early identification of both hazards and limitations are key factors in risk management. Once risk factors are identified, countermeasures can be established to mitigate the risk or buffers can be put into place to reduce the potential negative impact on people and organizational accomplishment. Identified individual limitations can be overcome through skill development designed to improve coping abilities. Most traditional stress management programs function at this level of health.

Surveillance strategies

Measurement tools at this level of health focus on early identification of potential hazards and risks. Proactive assessment and periodic monitoring forestall the onset of insidious disease processes. Continuing observation, or alertness, is warranted to identify potentially unexpected or gradually increasing problems or indicators of distress.

Organizational Health Risk Assessments (OHRAs)

Organizational Health Risk Assessments are similar to individual HRAs with the exception that they are designed for early identification of risks presenting to the entire organization as opposed to one individual. Periodic OHRAs help us to keep our finger on the pulse of the organization. Data from the OHRA identify potential environmental factors for modification, individual resources for enhancement, and potential areas of strain for support and treatment. A number of processes are used, depending on the base and the nature of the needs assessed there.

One such assessment is the DOD Survey of Health Related Behavior (SHB). The SHB (Bray *et al.* 1995) is a large-scale assessment of military members from all four services – the Air Force, Army, Navy, and Marine Corps of the United States. It was originally designed to examine substance use and health behaviors among military personnel. The sixth in a series of anonymous, self-administered questionnaires conducted worldwide since 1980, the 1995 repetition added questions related to stress and coping for the first time. The sample size was 16,193 active duty military personnel and the overall response rate was 79 percent.

Data reflect that military members perceive stress as emanating from both work and family or personal concerns. However, higher levels of stress were related to work issues, as shown in Figure 7.2. Associated with decreased personnel and increased responsibilities, increased workload was identified by over 16 percent as producing a high amount of stress. Increased deployments associated with a changing mission and

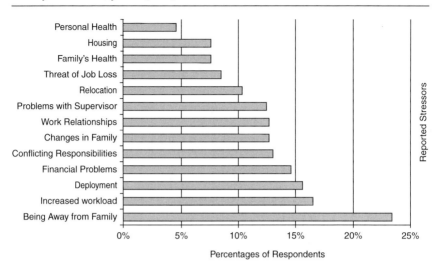

Figure 7.2   Percentages of respondents reporting "a great deal" or a "fairly large amount" of stress in the past 12 months. Adapted from the DOD *Survey of Health Behaviors* (Bray *et al.* 1995)

structure take people increasingly away from home. As a result, 23 percent of military personnel report high stress associated with being away from home and nearly 16 percent echo concerns of stress associated specifically with deployments.

High levels of stress appear also to be associated with negative work performance. Personnel who report high levels of stress are more likely to report that they arrive to work late, leave work early, or are absent from work altogether. They also report a higher incidence of on-the-job accidents. While at work, they tend to report a level of work performance lower than what they perceive to be normal for them.

On the positive side, effective coping strategies were widely used. The most commonly used strategies for coping with stress, as shown in Figure 7.3, include active planning or problem-solving, seeking social support through talking to a friend or family member, and positive self-care behaviors such as exercise or sports. Unfortunately, approximately 25 percent commonly used alcohol to cope with stress, daily pressures and feelings of depression. One in six reported they engaged in heavy or binge drinking – 17 percent compared with 12 percent on the part of their civilian counterparts. Those respondents aged 18–25 years were more likely to smoke as well – 39 percent compared to a national average of 35 percent. Those who reported heavy use of alcohol were also more likely to perceive a great deal of stress at work and at home and to experience more days when their mental healthcare was not good.

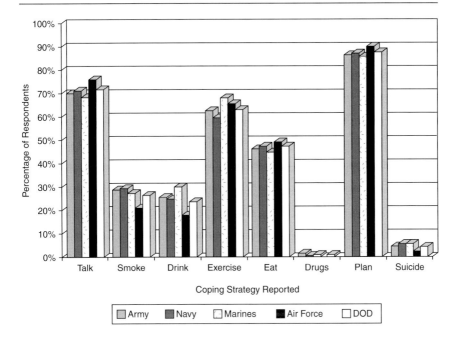

Figure 7.3 Percentages of personnel who "frequently" or "sometimes" engage in behavior when they feel pressured, stressed, depressed, or anxious. Adapted from the DOD *Survey of Health Behaviors* (Bray *et al.* 1995)

Aircrew Safety Inventory

The Aircrew Safety Inventory was designed to identify risks specific to the operational flying mission. This instrument allowed for both quantitative and qualitative responses on a number of issues previously identified in focus groups of aircrew members. The questionnaire was administered several times each year from 1996 through 1998 to a group of squadron-level safety officers who were also active pilots and navigators. Aggregate quantitative data from those administrations reflect a high level of attention to safety practices and strong values associated with and adherence to high standards of performance among this population. However, qualitative responses echo the concerns identified in the SHB that additional duties and deployments were the biggest problems faced by aviators. Family issues were perceived to be a result of work-related issues rather than the reverse. Increased time at work or away from home on deployments led to increased concern about care for family members while the military member was away. As services declined with overall resource reductions, the eroding of family services and family health was a serious concern for military members. Aircrew members also reported a concern that the original contract they perceived between the military service and the American public was being changed in the middle of their service, without their input or control. Changes involved decreased healthcare benefits and diminished retirement benefits. As the culture became more civilianized, they perceived the traditional culture of camaraderie beginning to erode as pilots working long hours chose to spend time at home with families rather than with their squadron mates, decreasing the positive collaborative relationships that characterized for years the health of the flying squadrons. Pay inequities have been a long-term characteristic of military service, but are often justified by service members through a dedication to public service and patriotic goals. However, changes in benefits, along with increased civilianization and outsourcing, led to a perceived change in the value the public placed on this service and magnified the pay inequities among lower ranking troops, tempting skilled military members to seek civilian employment.

This information fed into a larger task force on change and workload resulting in broad structural and process changes being made to mitigate identified risk factors. To reduce the burden of frequent and lengthy deployments, the service is being reorganized into an Expeditionary Air Force (EAF) structure with predictable deployments of predetermined teams. The acting secretary of the Air Force has also proposed a series of pay raises to attempt to decrease the wide gap between military and civilian pay and to restore retirement benefits. Access to healthcare for dependants and retirees is also being addressed (Briggs 1999).

Preventive Health Assessments (PHAs)

Preventive Health Assessments focus on the individual within the organization and consist of annual physical exam for all personnel. These examinations include a fitness test and completion of an individual HRA designed for early identification of individual health risks and high-risk behaviors.

Secondary prevention strategies

Health and Wellness Centers (HAWCs)

Health promotion activities are conducted through HAWCs located at every base. The HAWC program centers around seven components of wellness: stress management, tobacco cessation, substance abuse prevention, nutrition, fitness, and heart/cancer disease prevention. All active duty personnel are monitored on weight and fitness measures to ensure individual levels of health and fitness are sufficient to match the demands of high levels of physical and psychological stress in the military environment.

To integrate the efforts of multiple disciplines involved in wellness activities, a base Wellness Council meets regularly for planning and monitoring of goals and outcomes. The wing commander chairs the Wellness Council, making health promotion a leadership responsibility.

Suicide risk reduction

Personnel in the Armed Forces represent a highly select group of individuals who are physically and psychologically fit by necessity given the variety of demands to which they are exposed. It is not surprising that death rates overall are quite low. Most mortality results from accidental death, as is typical of most young adults. Suicide rates within the Air Force have been consistently lower than in the civilian population, as would be expected in a highly select group. Nevertheless, deaths from suicide began to increase in 1991, surpassing disease in 1995 to become the second leading cause of death among Air Force personnel. This situation triggered a concern that the stress of change and drawdown may have been increasing this risk. Even with the increase in rates, suicide remained a low base rate event. However, for every completed suicide in the Air Force, approximately six non-fatal acts of self-injury are reported. In addition, the death of a colleague can have a crippling effect on the functioning of a squadron or work unit that can linger for some time. The tragic losses associated with suicide are also compounded by the distress of family members. Additionally, the Centers for Disease Control contend that suicide and suicide attempts represent only the tip of an iceberg composed of poor coping, poor work performance, poor interpersonal relationships, and significant emotional distress as well as increased susceptibility to physical illness. For all these reasons, an intensive effort of risk identification and prevention was initiated by the Chief of Staff of the Air Force in collaboration with the Air Force Surgeon General. They brought together a cross-functional team of 75 professionals from around the globe to develop and implement a prevention program.

This Suicide Prevention Team amassed a significant amount of data related to suicides among Air Force personnel as a starting place for risk control and management. Demographically, the most likely people to complete suicide were Caucasian males between the ages of 24 and 44 in higher enlisted ranks who were either separated or divorced. Individuals who attempted, but did not complete, suicides tended to be younger (age 17–24) and lower ranking enlisted females who were also either separated or divorced. The most significant psychosocial risk factor for completed suicides was the presence of marital or relationship discord. Individuals who experienced legal problems or were under investigation were also at high risk, regardless of problem severity or guilt. Depression and substance abuse, most usually alcohol, were also identified as risk factors. More importantly, these factors were not mutually exclusive. Individuals who experienced multiple risk factors were at higher risk. These individuals often faced multiple new problems and did not have the specific skills needed to adapt to the situation or their coping skills had deteriorated and could not be applied at the time they were needed. Social support, either informal or professional, appeared to be lacking.

Programs of prevention focused on improved surveillance and multi-faceted interventions. Surveillance systems were designed to collect information both on risk factors and on prevention efforts. Data were standardized across the Air Force active, guard, and reserve forces and are now collected electronically in a central repository. This information is analyzed regularly to identify trends and results are distributed to intervention agents.

A comprehensive program of intervention was initiated which included increasing accurate awareness and information along with education and training in recognizing risk factors and getting appropriate assistance. Policy was developed that explicitly makes it a supervisory responsibility to provide extra support to individuals under investigation or in legal trouble. Special arrangements were made to increase access to confidential counseling for those individuals, to provide them a safe haven in which to discuss their concerns. Suicide risk reduction training is now provided annually for all Air Force personnel. This training is intended to help personnel identify individuals who are struggling and may be at increased risk for a suicide attempt. Additionally, the training provides instruction on what can be done to help, including how to make appropriate referrals to helping agencies.

Increased collaboration among support agencies involved in preventive services was achieved through establishing an Integrated Delivery System (IDS) at each base to provide a comprehensive safety net so that individuals at risk did not fall through cracks in the system. Organization-specific risk factors can be identified through the IDS using a variety of risk factor surveys administered in consultation with base leadership. Finally, post-intervention services were standardized including critical incident stress management procedures for victims and responders following a traumatic event, such as a suicide, with hopes of preventing further tragedy.

The efforts of the Suicide Prevention Team appear to be paying off. The suicide rate decreased 38 percent from 1995 to 1996 and an additional 25 percent from 1996 to 1997. The rate for 1998 is expected to finish at less than 10 per 100,000 active duty personnel, which is in the same range as the rates prior to 1990. The downward movement of suicide rates approaches significance ($p = 0.052$) suggesting a positive trend. There are a number of possible explanations for these results which need further evaluation; however, suicide has now declined to the third leading cause of death among Air Force personnel.

Healthcon Charlie

This is lowest level of health for continuously functioning organizations, absent extreme situations such as complete demise, disaster, or catastrophic events. Hazards are perceived as threats rather than either challenges or opportunities. The transaction of increased pressure with individual and organizational vulnerabilities produces a high level of risk. Dysfunctional, and, at times, counterproductive or destructive behavior may exist; organizational effectiveness is likely comprised. Assessment strategies focus on outcomes of the problematic situations or processes and monitoring of results of programs of intervention. Interventions, in turn, are designed to provide support, treatment, and rehabilitation to employees and to organizational units, to ease adjustment to change, and to reduce the probability of recurrence of similar problems in the future.

Data used for monitoring this level of health typically include morbidity and mortality rates, accidents and injuries, lost time, workers' compensation claims, and retention rates.

Tertiary prevention strategies

Safety investigations

High levels of sustained stress have a predictable impact on human performance. Innovation and problem-solving capability, communication, and teamwork are all decreased. Perceptions are distorted and decision-making is impeded. Ability to recall information and to apply that information in a novel context is also impaired. These performance decrements have significant implications for safety in the high-risk environment of aviation (Stokes and Kite 1994).

Human factors are identified as causal in approximately 75 percent of all aircraft accidents. That percentage holds true across military and civilian aviation and across international boundaries. Within the US Air Force, between 1989 and 1995, human factors were found causal in 74 percent of Class A aviation mishaps – which include those accidents in which there is loss of life and/or damage costs in excess of $1 million dollars. Of those factors, 65 percent were related to the individual and 58 percent were related to environmental human factors, with some expected overlap.

The Air Force has experienced an admirable aviation safety record and has led the way among the military services in low accident rates. Over the course of the past 20 years, accident rates have continually declined (see Figure 7.4). However, over the past few years, rates have seemed to plateau. Many advances have been made in equipment reliability and improved protective equipment and safety devices, but throughout this

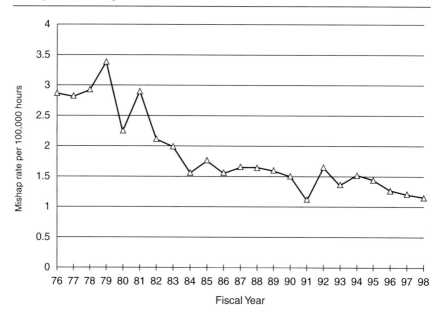

Figure 7.4 Air Force Class A Flight Mishap Rates: FY 76–FY 98. Courtesy of the United States Air Force Safety Center.

time, the percentage of mishaps related to human factors has remained fairly consistent. Therefore, the human factors arena appears to be an important area for intensive prevention efforts to continue to reduce mishap rates.

The Air Force Safety Center is responsible for prevention and investigation of aircraft accidents. Three strategies have been put into place for this purpose. The first is the Operational Risk Management (ORM) process, as described above. The Safety Center seeks to institutionalize a risk management culture to reduce risk at the point of mishap with each individual and each increasing level of supervision. The second is through the OHA process. OHAs are designed to identify strengths and weaknesses of organizations before the mishap occurs, mitigating the risk through proactive change.

The last is through investigation of mishaps after they occur. Thorough safety investigations are conducted for every aviation mishap. A board of officers and technical experts convenes for the sole purpose of identifying causes of the mishap and making recommendations for correction of identified problems to prevent a similar accident from happening in the future. These recommendations are then tracked by the Safety Center to ensure they are not forgotten. Given the emphasis on the human element in the majority of mishaps, human factors consultants are now routinely included in the safety investigation process. Aviation psychologists and physiologists join the flight surgeon, the traditionally assigned medical representative to the board, in thoroughly investigating all possible human factors – individual and organizational. The lessons learned from these investigations are then carried forward in safety training and procedural changes to assist in reducing injury and mortality through aviation mishaps.

Mental Health in Primary Care

Mental health problems have both a direct and an indirect impact on workplace productivity for employees who experience the problems as well as their colleagues. Costs of mental health problems incurred by individuals, families, communities, and society at large are measured in terms of decreased earning potential, role functioning, and quality of life. Of the health problems surveyed, the SHB found that depression had the most consistently significant effect on ability to function well on the job (Camilin *et al.* 1997). Results take from the Air Force Health Enrollment Assessment Review (HEAR 1.0, Region 6, CY 96) indicate rates as high as 24 percent for personnel feeling depressed, down, or hopeless, and 28 percent being bothered by "nerves" or feeling on edge. Yet only about 6 percent of HEAR respondents report seeing a healthcare professional about the symptoms (Lombard and Lombard 1997).

Reports of undiagnosed and untreated psychological and behavioral problems in the healthcare system abound. A series of studies found that 15 to 40 percent of all primary care patients have a diagnosable mental health disorder; 20 percent of primary care complaints are mental health related; and 50 to 75 percent of mental health problems that pass through primary care clinics are undiagnosed and untreated (Barret *et al.* 1988; Ormel *et al.* 1991; Ryden *et al.* 1992). A 20-year review of healthcare visits by the Kaiser health organization found that in 60 percent of all visits, no physical problem could be diagnosed. In an additional 20 to 30 percent of visits, the physical illness had a stress-related component (Keita, cited in Piltch 1993: 12). Similarly, a study within the Aetna health system found that 96 percent of all primary care visits had a psychological component (Holder and Blose 1987). Within the military healthcare system, Rundell (1992) reported that affective disorders are 10 times more common among medical patients than the general population. In addition, 70 percent of patients with an affective disorder are seen by at least 10 physicians before the mental health disorder is either referred or treated. High utilizers of healthcare also have a 20 to 80 times higher prevalence of psychological problems than the general population. Overall healthcare utilization tends to decrease following appropriate treatment, including reductions ranging from 5 to 88 percent subsequent to psychological or behavioral treatment, and from 26 to 69 percent subsequent to substance abuse treatment (Follete and Cummings 1987; Rundell 1992). Every trip to the healthcare provider represents an increase in lost time to the organization, not to mention a decrease in quality of life for the individual.

The Mental Health in Primary Care Project was designed to begin the process of addressing these problems. The three primary goals were to increase identification of behavioral health problems by the primary care team; to increase treatment of behavioral health problems within the primary care environment; and to identify and treat high utilizers of healthcare services. By providing behavioral health practitioners in a primary care setting, patients are able to avoid trips to specialty clinics, decreasing the stigma and inconvenience of obtaining mental health treatment and increasing likelihood of treatment attendance. The addition of psychologists to the primary care clinic also increases collaboration among the team of healthcare providers, increasing understanding of when and how to refer a likely psychological or behavioral problem. As a result, patients are referred more rapidly, receive treatment appropriate to their problem early in the disease process, and are returned to full performance capabilities more quickly. Patients experience less frustration in navigating the healthcare system and the organization, as a whole, benefits. This program also fulfills goals of the Suicide Risk Reduction program by providing an opportunity for early mental health intervention, increased access to mental health services, and decreased "missed opportunities" for intervention for individuals who are potentially at-risk (Lombard and Lombard 1997). This service is currently established as a demonstration program. Project staff report that, prior to initiation of the program, untreated mental health problems cost the clinic over $400,000 in excess utilization, a portion of which they intended to save through the program services. The project is now in its first year of operations; outcome data are expected after 18 months.

Conclusions

Change is a key feature of the new organizational reality. Individuals and organizations need the ability to adapt and flex with change if they are to maintain vibrant health. When change is viewed as an opportunity rather than a threat, the silver lining in changing times can be seen by those who seize opportunities as they arise. Leveraging lessons learned in individual health may help organizational leaders and change agents optimize organizational health during times of challenge and change so as to nurture growth rather than trigger decline. Healthy, productive work demands that people give their best and in the process of doing so receive the support and encouragement to invigorate them to achieve their full potential. To expect excellence in the midst of dramatic change may seem counterintuitive, yet it is reasonable in healthy, supportive systems. The US Air Force is growing into the future by implementing preventive organizational health strategies at multiple levels aimed at enhancing vitality while mitigating against health risk factors. However, organizational health is never an end in itself. The goal is the process. Programs vary as times and situations change. The human capital dividends resulting from continually nurturing both individual well-being and organizational vitality promise to be well worth the investment.

Note

1   The views presented are those of the authors and do not necessarily reflect the official policies or position of the United States Air Force or Department of Defense. The authors would like to thank Phillip W. Steely for his helpful comments on an earlier draft of this chapter.

References

Adkins, J. A. (1999) Promoting organizational health: The evolving practice of occupational health psychology, *Professional Psychology: Research and Practice* 30: 129–37.

Adkins, J. A. (1998) Base closure: A case study in occupational stress and organizational decline, in M. K. Gowing, J. D. Kraft and J. C. Quick (eds) *The New Organizational Reality: Downsizing, Restructuring, and Revitalization* (pp. 111–42), Washington, DC: American Psychological Association.

Adkins, J. A. (1996) Organizational health: An organizational systems perspective, in D. White (ed.) *Proceedings of the Third Biennial International Conference on Advances in Management (Boston, MA)* 3: 100.

Barret, J. E., Barret, K. A., Oxman, P. E. and Gerber, D. D. (1988) The prevention of psychological disorders in primary care practice, *Archives of General Psychiatry* 45: 1100–6.

Bray, R. M., Kroutil, L. A., Wheeless, S. C., Marsden, M. E., Bailey, S. L., Fairbank, J. A. and Harford, T. C. (1995) *1995 DOD Survey of Health-related Behaviors Among Military Personnel* (Report No. RTI/601 9/06-FR), Research Triangle Park, NC: Research Triangle Institute.

Briggs, M. (1999) Secretary details Air Force priorities, *Air Force News Service*, January 5.

Camilin, C. S., Fairbank, J. A., Bray, R. M., Wheeless, S. C. and Dunteman, G. H. (1997) *Does Stress Differentially Affect the Work Performance of Military Women and Men?* Paper presented at the 125th annual meeting of the American Public Healthcare Association at Indianapolis, IN.

Cannon, W. B. (1929) *Bodily Changes in Pain, Hunger, Fear and Rage*, New York: D. Appleton-Century. (Original work published 1915.)

Cannon, W. B. (1932) *The Wisdom of the Body*, New York: W. W. Norton.

Cannon, W. B. (1935) Stresses and strains of homeostasis, *The American Journal of the Medical Sciences* 189: 1–14.

Cigrang, J. A., Carbone, E. G. and Todd, S. L. (in press) Mental health attrition in Air Force basic training, *Military Medicine*.

Cliatt, S. R. and Stanley, G. A. (1994) *A case study of the base-closure community initial redevelopment process* (AFIT/GLM/LAL/94S-7). Unpublished master's thesis, Air Force Institute of Technology, Wright Patterson Air Force Base, OH.

Cohen, W. S. (1985) Health promotion in the workplace: A prescription for good health, *American Psychologist* 40: 213–16.

Follete, W. and Cummings, N. (1987) Psychiatric services and medical utilization in a prepaid health plan setting, *Medical Care* 5: 15–35.

Foss, L. and Rothenberg, K. (1987) *The Second Medical Revolution: From Biomedical to Infomedical*, Boston, MA: New Science Library.

Frese, M. (1997) Dynamic self-reliance: An important concept for work in the twenty-first century, in C. L. Cooper and S. E. Jackson (eds) *Creating Tomorrow's Organizations: A Handbook for Future Research in Organizational Behavior* (pp. 399–416), Chichester, UK: John Wiley.

Gowing, M., Kraft, J. and Quick, J. C. (1998) *The New Organizational Reality: Downsizing, Restructuring, and Revitalization*, Washington, DC: American Psychological Association.

Hickox, K. (1994) Content and competitive, *Airman* January: 31–3.

Holder, H. and Blose, J. (1987) Changes in healthcare costs associated with mental health treatment, *Hospital and Community Psychiatry* 38: 1070–5.

King, R. E. (in press) *Aerospace Clinical Psychology*, Brookfield, VT: Ashgate Publishing.

King, R. E. and Flynn, C. F. (1995) Defining and measuring the "right stuff:" Neuropsychiatrically Enhanced Flight Screening (N-EFS), *Aviation, Space, and Environmental Medicine* 66: 951–6.

Kuper, S. and Barlow, C. (1998) Improving logistics process reengineering, *Leading Edge* 40(7): 15.

Last, J. M. and Wallace, R. B. (eds) (1992) *Public Health and Preventive Medicine* (13th edn) Norwalk, CN: Appleton and Lange.

Lombard, T. and Lombard, D. (1997) *Mental Health in Primary Care*. Presentation to the Air Force Surgeon General, Washington, DC.

Nelson, D. L., Quick, J. C. and Quick, J. D. (1989) Corporate warfare: Preventing combat stress and battle fatigue, *Organizational Dynamics* 18: 65–79.

Ordin, D. L. (1992) Surveillance, monitoring, and screening in occupational health, in J. Last and R. B. Wallace (eds) *Public Health and Preventive Medicine* (13th edn, pp. 551–8), Norwalk, CT: Appleton and Lange.

Ormel, J., Koeter, M. W. J., VandenBrink, W. and Van de Willege, R. (1991) Management and course of anxiety and depression in general practice, *Archives of General Psychiatry* 48: 700–6.

Piltch, C. A. (1993) Mental health and the workplace: Some emerging policy issues, in *Policy in Perspective* (pp. 2–13), Washington, DC: Mental Health Policy Resource Center.

Quick, J. C. (1999a) Occupational health psychology: Historical roots and future directions, *Health Psychology* 18: 82–8.

Quick, J. C. (1999b) Occupational health psychology: The convergence of health and clinical psychology with public health and preventive medicine in an organizational context, *Professional Psychology: Research and Practice* 30: 123–8.

Quick, J. C., Quick, J. D., Nelson, D. L. and Hurrell, J. J. Jr (1997) *Preventive Stress Management in Organizations*, Washington, DC: American Psychological Association. (Original work published in 1984 by J. C. Quick and J. D. Quick.)

Rundell, J. R. (1992) *The Impact of Psychiatric Consultation on Medical/Surgical Morbidity*. The William C. Porter lecture at the annual meeting of the Association of Military Surgeons of the United States, Nashville, TN.

Ryden, P., Redman, F., Sanson-Fisher, R.W. and Reid, A. (1992) Determination of alcohol-related problems in general practice, *Journal of Studies on Alcohol* 53: 197–202.

Sauter, S. L., Murphy, L. R. and Hurrell, J. J. (1990) Prevention of work-related psychological disorders: A national strategy proposed by the National Institute for Occupational Safety and Health (NIOSH), *American Psychologist* 45: 1146–58.

Shrivastava, P. (1987) *Bhopal: Anatomy of a Crisis*, Cambridge, MA: Ballinger.

Selye, H. (1973) Evolution of the stress concept, *American Scientist* 61: 692–9.

Selye, H. (1976) *Stress in Health and Disease*, Boston: Butterworths.

Staw, B. M., Sandelands, L. E. and Dutton, J. E. (1981) Threat-rigidity effects in organizational behavior: A multilevel analysis, *Administrative Science Quarterly* 26: 501–24.

Steely, P. W. (1999) Expect excellence, *Kelly Observer* 34(2): 4.

Stokes, A. and Kite, K. (1994) *Flight Stress: Stress, Fatigue, and Performance in Aviation*, Brookfield, VT: Ashgate Publishing.

Teasdale, E. L. and McKeown, S. (1994) Managing stress at work: The ICI-Zeneca Pharmaceuticals experience 1986–1993, in C. L. Cooper and S. Williams (eds) *Creating Healthy Work Organizations* (pp. 133–65), Chichester, UK: Wiley.

Weick, K. (1987) Organizational culture as a source of high reliability, *California Management Review* 29: 112–27.

Wellins, S. and Murphy, J. S. (1995) Reengineering: Plug into the human factor, *Training and Development* 1: 33–41.

Chapter 8

Individual and organizational activities to build better health

W. C. Schmidt, L. Welch, and M. G. Wilson

INTRODUCTION

The healthy work organization concept centers on the premise that organizations that foster employee health and well-being are also profitable and competitive in the marketplace. The concept recognizes that work can have a significant effect on employee commitment, satisfaction, and health which, in turn, impact productivity and the effectiveness of the organization (Williams 1994; Wilson in press). Healthy work organization represents the concentration of thinking and research from occupational safety and health management, human resources and organizational development, occupational stress, and worksite health promotion. These diverse disciplines are uniform in identifying convergent themes of healthy work organizations which include:

- the increased importance of organizational factors in the work/health relationship
- the need for organizational-level action in promoting positive changes
- the need for modification of the traditional employer/employee relationship by increasing opportunities for employee involvement and input (DeJoy *et al.* 1997).

The operationalization of healthy work organization is based on two assumptions. The first assumption is that it is possible to identify a set of job and organizational factors that characterize the healthy work organization. DeJoy and colleagues (in press and 1997) have proposed a model of healthy work organization that includes 13 factors or dimensions. These dimensions include workload, autonomy/control, role clarity, job content, work scheduling, environmental conditions, organizational support, participation and worker involvement, feedback and communication, job security, advancement and continuous learning opportunities, equitable pay and benefits, and flexible work arrangements. The dimensions impact employee perceptions and expectations about the organization and their work which include job satisfaction, organizational commitment, and stress symptoms (both physical and emotional distress). The workers' perceptions and expectations in turn impact organizational effectiveness.

The second assumption is that creating this type of workplace should result in healthier, safer, and more productive workers, the payoff of which would be increased profitability and competitiveness in the marketplace (Murphy 1995). Neither of these assumptions has been empirically tested. However, that doesn't mean that organizations haven't been active in attempting to promote a healthy workplace. The purpose of this chapter is to describe how one company has integrated healthy work organization concepts throughout its operations and how it continues to strive to foster emotional and physical health and well-being of its associates and the communities in which they reside.

Organizational background

Founded in 1978, the Home Depot is the world's largest home improvement retailer and ranks among the 10 largest retailers in the United States, with fiscal 1997 sales of $24.2 billion. At the close of fiscal 1997, the company was operating 624 stores, including 587 Home Depot stores and five EXPO design center stores in the US and 32 Home Depot stores in Canada, as well as wholly owned subsidiaries Maintenance Warehouse and National Blind and Wallpaper Factory. In addition, the Home Depot recently expanded into Puerto Rico and into South America with stores in Chile and Argentina. The company employed approximately 125,000 associates at the end of fiscal 1997 and was recently named America's most admired specialty retailer by *Fortune* magazine for the sixth consecutive year. The company has been publicly held since 1981.

The Home Depot's business strategy is to offer a broad assortment of high quality merchandise at low "day-in, day-out" warehouse prices and provide exceptional customer service through highly trained and knowledgeable associates. With the exception of the EXPO Design Center stores, the Home Depot stores are isomorphic with each store stocking approximately 40,000 to 50,000 different kinds of building materials, home improvement supplies, and lawn and garden products. Home Depot customers include "do-it-yourselfers" as well as professional business customers, including home improvement contractors, tradespeople, and building maintenance professionals. Expo Design Center stores provide products and services primarily related to design and renovation projects. Maintenance Warehouse offers facilities maintenance and repair products, and National Blind and Wallpaper Factory offers wallpaper and custom window treatments. Products from the latter two subsidiaries are provided via direct mail.

Organizational culture

From the very beginning, Home Depot's founders, Bernie Marcus and Arthur Blank, believed in doing the right thing. By taking care of its associates and the communities in which they reside, the company sought to achieve its goal of becoming a socially responsible company. The core of the company's success has undoubtedly come from a strong belief in its business values, and these values are beliefs that do not change over time; they serve to guide day-to-day decisions and actions. The Home Depot's own values include creating shareholder value, taking care of employees, providing excellent customer service, fostering an entrepreneurial spirit, having respect for all people, building strong relationships, trying to do the right thing, and giving back to the community. The Home Depot strongly believes that the values are inextricably linked to their bottom-line.

The Home Depot's efforts to build a socially responsible company begin with taking care of its employees and continue with efforts to give back to the community. The organization offers a variety of services, benefits, and financial programs for its employees. Employee services range from adoption assistance and elder care assistance to health promotion programs; benefit programs range from disability insurance to healthcare coverage; and financial programs range from an employee stock purchase plan to a matching gift program. Many of these programs are detailed later in the chapter.

When it comes to impacting communities, being a leading retailer in the home improvement industry lends itself to providing service in a variety of ways. The organization accomplishes this through their Team Depot program. The Team Depot program provides donations and manpower to help community-based non-profit organizations provide affordable housing, assist youth at risk, and protect the environment. Associates get the opportunity to impact their community and feel like they are part of an important team. In fact, the Home Depot and its associates have donated time and/or resources to over 300 organizations across the United States and Canada. In addition, the Home Depot has been innovative in establishing environmental programs that are designed to find environmentally friendly products for home use, provide recycling services for citizens and building professionals, and support the work of non-profit organizations around the globe to further environmental education and research. The following discussion will first detail how the organization supports the health and well-being of its associates through individual and organizational-level activities which are primarily directed through the Building Better Health program and then discuss how it serves the community through the Team Depot program.

BUILDING BETTER HEALTH

Background

The primary program designed to support associate health and well-being is the Building Better Health program. This health promotion program was informally created in 1982 when associates in the corporate office started inquiring about the availability of aerobics classes. Since there was no onsite fitness center, these early aerobic classes were taught by a volunteer in a large storage area. Two years later, the organization contracted to use a fitness center in a nearby apartment complex. As interest grew, the company president created a team of associates to organize a more official health promotion program. The team visited and researched other corporate health programs and finally interviewed and hired the first health promotion coordinator for the company. By this time, the corporate headquarters had relocated to a larger office building that had a fitness center. It was out of this facility and an adjoining office that programs were offered to associates. During that same year, a contest was held among associates to identify a recognizable name for the program. The name that was overwhelmingly selected was Building Better Health – very appropriate for a home improvement company. Two years later, the Building Better Health program was expanded beyond the corporate headquarters to all the stores. As the number as stores increased and the corporate headquarters staff grew to support those stores, additional staff were brought in to meet the demand. This included health promotion professionals at headquarters and in the divisional offices. Eventually the Home Depot constructed an office complex for their company headquarters that included a 24,000 square foot fitness facility, basketball court, and outdoor walking trail. In addition, each divisional office across the United States and Canada has been built with a small fitness center with the exception of one, which has a fitness facility across the street.

Mission and goals

Building Better Health's mission is to promote health and wellness to all associates and their families, enabling them to experience a higher quality of life. As such, the Building Better Health program exists for the company associates; it is intended to show them that the company cares. One primary goal of Building Better Health is to create a company culture that breeds healthy, productive associates. According to the president of the company, "Building Better Health is part of our culture. It offers an opportunity for all of our associates to enjoy a higher quality of life. After all, our associates are our most important resource and we are here to assist them in taking care of themselves." A company-wide survey of associates found that 92 percent of respondents thought Building Better Health was important to the Home Depot culture.

A second goal is to reduce the risk of developing a serious illness or disease through primary prevention programs. This may include everything from smoking cessation programs to annual influenza immunizations. Of particular interest are programs designed to help associates moderate levels of stress. The tremendous success and growth of the organization has placed a strain on associates to integrate a substantial number of new associates into the organization while maintaining the quality of services and programs for current associates and customers. The most recent random, cross-sectional survey of associates found that 42 percent of respondents indicated they experienced high to very high levels of work stress while 35 percent reported moderate levels of work stress.

A third goal is to identify associates with the highest risk and try to reduce that risk through secondary prevention programs. These programs consist of blood pressure, cholesterol, blood sugar, depression, and skin cancer screenings, mammography, and alcohol and drug treatment. Screenings are a key component of Building Better Health services and are conducted quarterly in the stores and corporate headquarters. The most recent company data indicate that 91 percent of associates have had their blood pressure checked and 70 percent have had their cholesterol checked within the past year.

Building Better Health structure

At every Home Depot site there is a wellness representative or a wellness advisory committee. These associates make Building Better Health a reality. Wellness representatives are associates employed at a store that volunteer to help their fellow associates learn more about health and taking care of themselves. Wellness representatives might come from a variety of backgrounds or departments within the store. They receive training on how to promote health and wellness to all associates through awareness, education, assessment, and intervention programs. They receive two hours a week on the clock to plan, organize, and implement programs. Wellness representatives plan and organize the health promotion activities at the store level which may range from health education programs on a variety of topics to local walks/runs or blood drives. The wellness advisory committee is a group of interested associates who help plan, publicize, and create excitement about the upcoming activities.

Wellness representatives meet quarterly with their district wellness captain to share ideas and resources. The wellness captain is one volunteer associate within each district (five to seven stores) that is very knowledgeable about Building Better Health activities as a result of extensive training and/or experience. In addition, wellness representatives receive quarterly packets of information, from Building Better Health staff at corporate headquarters, on different health topics to support their planning and activities. The kits include guidance on planning and implementing health promotion activities and a variety of support materials for the activities. Wellness representatives are encouraged to work with community-based non-profit agencies such as the American Heart Association or the American Cancer Society to help them conduct various activities and/or provide additional support materials. Each store has a Building Better Health budget that is directed by the wellness representative and can be used for materials, screenings, programs, and incentives.

At the next level is a Building Better Health/Team Depot Coordinator. The coordinators are based in the Divisional Offices and report to the Manager of Building Better Health, who is located at corporate headquarters. The coordinators' main responsibility is to support the division stores and offices through training and communication. The Manager of Building Better Health is responsible for all activities and services being offered to associates through Building Better Health. The manager coordinates personnel, budgets, long-range planning, contracts with outside organizations, and is the chief liaison with other departments within the company.

PROGRAMS AND ACTIVITIES DESIGNED TO CREATE HEALTHY WORK ORGANIZATIONS

Building Better Health activities and programs

Prenatal education

Building Better Health promotes the Dependent Care Connection prenatal kit and the L'il Appleseed prenatal education program. All pregnant associates and spouses receive prenatal education packets with information on nutrition, exercise, stress, and pregnancy. Wellness representatives promote these programs to their associates. In addition, the corporate office offers quarterly lactation classes for all associates and spouses as well as a lactation room in the wellness center for mothers to express their milk. The first year it was implemented, the program reduced premature birth claims by 57 percent.

Quit smoking program and $50 quit smoking benefit

This program provides one-on-one consultation and support with a quit smoking specialist, as well as brochures and education packets to help a smoker quit once and for all. The organization partially reimburses for the program through the health insurance plan. The Home Depot has been a smoke-free workplace since 1992. The company does allow smoking on company grounds in designated areas outside the stores and offices.

Health promotion programs

A variety of health promotion programs are offered to all associates throughout the year. Each quarter centers on a theme that drives the programmatic efforts. Previously conducted programs include nutrition, weight control, fitness, stress management, back health, hypertension, hypercholesterolemia, depression, breast cancer, self-care, HIV/AIDS, and personal and family safety.

Screenings

In order to identify individuals at high risk for chronic disease or illness, screenings are offered on a quarterly basis to associates in the stores as well as in the corporate office. Types of screenings conducted include blood pressure, cholesterol, blood sugar, depression, percent body fat, flexibility, and cancer. Individuals identified as high risk receive educational counseling and referral to their physician for follow-up.

Water bottles for back belts

Water bottle holders are designed to offer an immediate supply of water for the associates to help avoid dehydration and heat exhaustion. They are distributed by the wellness representatives and/or store managers. The back belts protect and help support associates lifting heavy items, while the water bottles help to keep associates well hydrated particularly during the summer months. Ultimately these two programs tie back to healthier and more productive associates through a reduction of acute low back injuries.

Healthy vending machines

A selection of healthy items are required to be offered in the vending machines located in employee break rooms. The contract with a national vendor is reassessed annually in order to maintain a healthy selection of food options for associates. When associates have the opportunity to eat better, they reduce their risk of chronic diseases and have increased levels of energy for longer periods of time.

Cafeteria

This program works closely with the corporate office cafeteria to provide a variety of healthy meal selections including low fat, low sodium, low cholesterol, and high fiber items. An example of this is the Happy Heart club. This program was designed to encourage people to avoid fried selections and eat healthier selection. Associates receive a frequent healthy meal card. Each time they select the healthy selection of the day, the cashier stamps their card. After the tenth stamp, the associate receives $5 towards the next purchase of food as well as incentive points towards free items (e.g. polo shirts, sweat pants).

Registered dietitians and certified trainers

Two registered dietitians on staff are available for consultations on a variety of nutritional issues including diabetes, cholesterol, weight loss, and pregnancy. The program also has certified exercise trainers that provide fitness evaluations and exercise prescriptions to associates wanting to start an exercise program or modify their current program. Consultations are offered to all associates in person or over the phone.

National fitness center discounts

At the corporate and divisional offices fitness centers are available, and all associates, including those in surrounding stores, are eligible to join. In addition, Building Better Health offers nationwide corporate discount rates for area fitness centers in order to accommodate stores without fitness centers and offer similar programs to all associates throughout the company,

Fitness series

The Home Depot is a committed sponsor of community-wide fitness events. There are avid groups of walkers, runners, and cyclists who take to the streets as part of the fitness series, which benefits non-profit youth and health and human services organizations in dozens of locations in the United States and Canada. It also provides an opportunity for developing partnerships and team-building among associates who participate. Building Better Health sponsors over-40 races across North America benefitting various non-profit community organizations including the Boys and Girls Club, Walk for Troubled Youth, American Cancer Society, American Heart Association, and the Rainforest Alliance.

HIV/AIDS training program

Workplace HIV/AIDS training programs through the American Red Cross are offered for all managers. The program addresses legal and organizational as well as individual issues related to HIV+ employees. All managers and supervisors in stores two years or older have been trained.

Wellness champion recognition program

This recognition program awards special patches to associates who have made and met a health goal. For example, associates who quit smoking, lose a significant amount of weight, or run a race are awarded wellness champion patches. Nominations are accepted and patches are awarded quarterly. The purpose of this is to positively reinforce healthy behavior by publicly recognizing associates who have accomplished a health-related goal.

Other organizational activities and programs

The healthy work organization concept embraces the idea that healthy, happy workers are productive workers and productive workers create a healthy work organization. Much of the previous discussion has centered around programs and services designed to facilitate healthy workers. However, there are also a variety of programs and services that can contribute to the well-being of workers by supporting workers financially and professionally. An example of these programs and services includes the following.

Healthcare plans

In the United States, the Home Depot offers medical, vision care, and dental plans for all regular, full-time associates and qualified retirees through a preferred provider organization. The plans cover dependents as well as the associate, and includes well-child care, well-woman care, and prenatal education programs. Associates in Canada and South America are covered through their respective government's insurance plan.

Insurance plans

All regular, full-time associates qualify for life, accidental death and dismemberment, and long-term disability insurance plans. The organization provides basic coverage and the option to upgrade coverage if desired. Dependents are also eligible for coverage under the plans.

Financial plans

The organization offers a variety of financial plans to help associates prepare for the future. All associates with the organization for at least a year can take part in a 401(K) plan or an employee stock purchase plan. The employee stock purchase plan offers shares of the organization at a substantial discount from the current market price.

Child care program

Child care assistant programs are offered to all associates. This resource and referral program helps associates find, evaluate, and choose appropriate child care arrangements. A 10 percent tuition discount is also offered through agreements with two national child care agencies.

Adoption program

This program provides family care leave for eligible adoptive parents. The organization also offers reimbursement up to $3000 for eligible adoption expenses. The program offers referrals, consultation, and support before, during, and after the adoption process.

Elder care program

This offers resource and referral programs which provide information on aging, types of services available, referrals to community services for the elderly and helpful publications.

Employee assistance program (CARE program)

The CARE program offers resources and referrals for associates seeking counseling for individual, family, marital, credit, or substance abuse problems and legal aid. Initial assessment is made by telephone and individuals in need are referred to a qualified clinician in their area. The service is available 24 hours a day, seven days a week.

Family leave program

This offers unpaid family leave for up to three months that may be used by associates to care for an immediate family member. The program also provides extended family leave for up to an additional three months for eligible associates.

School match program

This functions as a resource and referral service that assists parents in choosing appropriate public or private schools for their children.

Relocation program

This program is designed for associates relocating to a different community. It provides resources and referrals to help find out about school districts and child care in the new community.

Executive physicals

The executive health program conducted in conjunction with a local university is designed to monitor and maintain optimum health for company officers. Comprehensive annual medical examinations include age-based screening and review of personal and family health histories. Unlike traditional physical examinations, the goal of the program is not just diagnostic, but is designed to provide officers with the tools and understanding to facilitate a healthy lifestyle.

Team Depot

Team Depot is the associate volunteer force linking the company to community needs through hands-on-service. Each store has a Team Depot representative who coordinates activities and associate participation. They are supported by a coordinator at the division level and a manager at corporate headquarters, similar to the Building Better Health structure. Ongoing Team Depot programs include Habitat for Humanity, Christmas in April, KaBOOM!, and the Disaster Relief project.

Habitat for Humanity

Associates volunteer their time and the organization donates materials to build affordable housing through the Habitat for Humanity program. This program builds houses for low income individuals unable to afford housing on their own. Currently, associates are working with 97 Habitat affiliates throughout the United States and Canada.

Christmas in April

The organization has established a relationship with the non-profit organization Christmas in April, to help rehabilitate housing for the elderly and disabled. In 1997, associates worked in 93 communities to mend leaky roofs, replace worn carpet, and fix broken staircases.

KaBOOM!

The Home Depot and its associates also work with KaBOOM!, an organization that creates a safe place for children in urban areas to play by assembling playgrounds. The company has committed to help KaBOOM! build 1000 playgrounds by the year 2000.

The Disaster Relief project

This project partners the organization with major relief agencies to provide aid to communities experiencing a natural disaster. Disaster Relief provides much needed supplies, such as water, building materials, and volunteer labor to assist those in need.

IMPACT OF ACTIVITIES AND PROGRAMS

Currently, the Home Depot is growing at a rate of 22 percent which means approximately 150 stores and 35,000 new associates will be added to the organization over the next fiscal year. In addition, as with any retail organization, there is a loss of workers due to turnover in stores that must be replaced. This phenomenal growth creates a number of challenges for the organization, not least of which is just monitoring participation in the various programs and services discussed above. In a rapidly changing corporate environment, conducting evaluations of the impact of programs and services becomes a goal that is desired but normally not achieved. Hence, many organizations in this position rely on subjective measures of the impact of their programs and services or sporadic evaluations of individual programs. This may shed light on one piece of the puzzle, but does not provide a complete understanding of the impact of programs and services on employee health and well-being or organizational effectiveness.

The Home Depot has undertaken an evaluation effort designed to examine the impact of individual programs and services and to provide longitudinal measures of changes in employee health and well-being and organizational effectiveness. The evaluation effort has been implemented through the Building Better Health program and incorporated many of the programs and services described above. The first component of the evaluation effort was the development of an evaluation plan designed to provide a comprehensive, longitudinal examination of employee health and well-being. This plan has two foci, fashioned to look at short-term impacts of specific programs and long-term outcomes of programs and services on associate health and well-being and the organization.

The short-term evaluation strategy was constructed to measure the effectiveness of specific programs on employee participation, satisfaction, and health risk behaviors. It was also intended to serve as a mechanism to conduct process evaluations to strengthen existing programs and services. The specific programs to be targeted for this evaluation strategy are determined annually at the beginning of the fiscal year and usually change from year to year. Previous evaluations have examined the Building Better Health communication structure, the educational packets used by the wellness representatives, and the prenatal care program.

The long-term strategy was intended to measure changes in the health and well-being of the Home Depot population over an extended period of time and to determine the effects of programs on absenteeism and healthcare costs. In order to measure changes in the health and well-being of the workforce over time, a cross-sectional survey has been administered annually to a random sample of associates throughout the company. This strategy was similar to that used by the United States government to build a picture of the health of the country by conducting annual surveys such as the Behavioral Risk Factor Surveillance Survey or National Health Interview Survey. Questions asked on the survey inquire about current health risk behaviors, participation in and satisfaction with a variety of programs and services, and employee perceptions and expectations about their work and the organization. In addition, a tracking system has been developed that was designed to measure changes in absenteeism and health-care costs over time. When an associate participates in a program, screening, annual physical, the fitness center, or the annual survey, the data gleaned from these activities are entered into a data base that will track the individual's participation and health-related behaviors. This data base can be matched with other data bases that track absenteeism, workers' compensation claims, healthcare claims, and recordable illnesses and injuries to better understand the relationship between these outcome variables, individual health and well-being, and organizational effectiveness.

The second component of the evaluation effort is an empirical test of the healthy work organization concept, its impact on the organization, and a process for intervening to create a healthy work organization. This pilot study is based on the healthy work organization model developed by DeJoy and colleagues (in press and 1997). The purposes of the study are to:

- determine what factors or variables influence the development and/or maintenance of a healthy work organization, and
- test a process for intervening to improve healthy work organization.

The intervention uses data-driven, problem-solving teams that combine elements from total quality management, worker involvement, and community engagement. Data collection includes quantitative and qualitative measures of healthy work organization dimensions, health and well-being, and financial performance. The intervention will be administered and data collected over a three-year period. The results of this pilot study will provide feedback to the organization on how well it is taking care of its employees and whether or not doing so impacts organizational effectiveness.

An additional evaluation study has been conducted to determine the effects of a change in back support use policy on the occurrence of work-related low back injuries among the Home Depot associates (Kraus *et al.* 1996). Over 36,000 associates participated in the study that measured working hours of exposure, back support use, intensity of materials-lifting requirements, and injury-related claims over a five-year period. The results indicated that mandatory implementation of a back support use policy significantly reduced the incidence of acute low back injuries. The Home Depot used this evaluation study to modify its back support use policy to facilitate use of back supports among its associates, decreasing the likelihood that they will be injured or disabled.

FUTURE DIRECTIONS

The Home Depot is convinced of the benefits of promoting health for all of its associates. Today, many businesses offer some form of health promotion programs to their employees and they do so primarily because they believe the benefit is worth the cost. In the future, the organization, through the Building Better Health program, wants to be able to quantify financial and health outcomes for its health promotion programs. The goal is to show a quarterly return on the investments made to the most popular programs like exercise, nutrition, smoking cessation, and stress management. In order to accomplish this, Building Better Health needs to revise and refine the current process for collecting and analyzing data. It has been said that worksite wellness is healthcare reform that works, and the Home Depot would like to show the impact the investment in wellness can have on the bottom line.

In addition, Building Better Health would like to further integrate health promotion with overall benefit and business strategies and work even better with our Benefits and Risk Management and Occupational Health and Safety departments to provide total health management. As previously discussed, the Home Depot is growing so rapidly that the bulk of the organization's time is spent keeping track of who has what benefits. It becomes difficult to take a proactive approach to health management when a large amount of time is spent reacting to continual change. Over the next decade, additional emphasis will be placed on primary prevention programs and services such as annual physicals, diabetes screenings, and influenza immunizations. These represent important avenues by which to identify individuals prior to the development of disease and aim to reduce costs over the long term. Consequently, a primary goal is to have the Building Better Health and Benefits and Risk Management departments work more efficiently and effectively to take better advantage of expertise on staff and programs available through the managed care organization to provide a more comprehensive health management program. This will allow the Home Depot to take care of its people throughout the spectrum of disease prevention and treatment which will ultimately lead to healthcare dollars saved and increased quality of life.

One final goal for the future will be to keep the Home Depot culture alive as the organization continues to grow and add new associates. One way to accomplish that will be to enhance communications from top to bottom about the company values and the ways the organization is helping to take care of its people. Associates cannot take action to better their health and well-being if they are unaware of the programs and activities in place to support the action. A second way to achieve that goal will be to empower associates to take better care of themselves and their families and to become more active in the operation of the organization. Associates need to feel that their health and well-being and the success of the organization is within their control and their actions will result in a better quality of life for themselves and a more successful organization. A third way to attain the goal will be to increase participation of all associates. Associates need to be active partners in the organization and their own health. Increased participation can be facilitated through increased communication and empowerment, all three of which go hand in hand.

There is a family feeling at the Home Depot that isn't found at many companies and the organization is striving to continue that. Reinforcing the company values through communication, empowerment, and participation and continuing to provide the best programs and services possible will enable the company's associates to live better, longer and more productive lives. As a company officer has said, "Our associates are more than associates – they're people and our extended family. We care about the overall person and want to enhance their careers and their lives."

References

DeJoy, D. M., Wilson, M. G., Griffin, C. S. and Peer, M. K. (1997) *Defining and Operationalizing the Concept of Healthy Work Organization*. (Final Report: PO0009638040) Cincinnati, OH: National Institute for Occupational Safety and Health.

DeJoy, D. M., Wilson, M. G., and Griffin-Blake, C. S. (in press) Healthy work organization, in W. Karwowski (ed.) *International Encyclopedia of Ergonomics and Human Factors*, Taylor & Francis.

Kraus, J. F., Brown, K. A., McArthur, D. L., Peek-Asa, C., Samaniego, L. and Kraus, C. (1996) Reduction of acute low back injuries by use of back supports, *International Journal of Occupational and Environmental Health* 2: 264–73.

Murphy, L. M. (1995) Characteristics of health work organizations. Symposium conducted at the American Psychological Association/National Institute for Occupational Safety and Health Conference, Work, Stress, and Health 1995: Creating Healthier Workplaces, Washington, DC.

Williams, S. (1994) Ways of creating healthy work organizations, in C. L. Cooper and S. Williams (eds) *Creating Healthy Work Organizations*, Chichester, UK: John Wiley.

Wilson, M. G. (in press) Health promotion in the workplace, in J. R. Rippe (ed.) *Lifestyle Medicine*, Cambridge, MA: Blackwell.

Chapter 9

Bringing health to life

*E. L. Teasdale, R. J. L. Heron, and
J. A. Tomenson*

INTRODUCTION

The current world of work and home presents great challenges to people and to organizations. The ability to balance often competing demands is a skill that leads to well-motivated and well-rounded people, capable of giving of their best in all walks of life.

The authors of this chapter work for AstraZeneca, which was established in April 1999. It is a company which invents, develops, manufactures, and markets products to improve human health, nutrition and quality of life around the world.

AstraZeneca (formed from the merger of Astra and Zeneca) is synonymous with successful innovation, together with wide-ranging manufacturing and commercial strengths. Headquartered in the UK, AstraZeneca operates internationally and sells in over 100 countries worldwide, combining its global capabilities with high quality customer relationships in local markets. As a company, the challenging nature of the working environment is recognized. The company operates globally in an increasingly competitive marketplace, with ever more rigorous demands from regulatory authorities. The importance of keeping staff motivated and maintaining their health and well-being is well recognized.

The work detailed in this chapter was carried out in Zeneca between 1987 and 1999. (Astra has also worked on similar lines.) The programs introduced were designed to encourage thought about life management, help individuals and managers to recognize signs of stress and provide routes of support. Recognizing mental health problems at an early stage and providing access to confidential support in a workplace setting usually helps to resolve problems before they create difficulties.

In this chapter "Health" and "Stress" are addressed as topics. The incidence and legal and financial impacts of stress-related illness are also discussed. A "Model of Well-Being" is displayed, one element of which focuses on training and, specifically, life management skills including stress management. In Zeneca, workshops on stress management have been held for more than 10 years: a study was recently undertaken on the benefits. The "Model of Well-Being" also highlights the importance of support services such as "CALM" (an in-house counseling and proactive, education provision in life skills). In the conclusion to this chapter, the current and future focus of occupational mental health is discussed.

SO WHAT IS "HEALTH"?

The World Health Organization defines healthcare as a state of complete physical, mental, social and spiritual well-being. The spiritual "angle" has nothing to do with religion, it merely relates to the need we all have to be treated as individuals and to grow and develop during our adult life. Many of the "physical" aspects of occupational healthcare are now well understood; for example dermatitis and noise-induced hearing loss. Over the last 10–20 years, the mental, social and spiritual side of life has become much more prominent and is probably more difficult to understand and manage.

   In the occupational or industrial setting, the emphasis on maintaining *mental* (as well as physical and social) well-being is essential to success. Occasional serious cases of mental illness must be recognized and managed appropriately. This should include the care of the individual who has a problem of substance abuse (e.g. alcohol or drugs of addiction), and the patient with a psychotic condition where prompt admission to hospital is required. In practice, however, the more common mental health problems encompass stress, and stress-related illnesses and states such as anxiety and depression, and their manifestation in the workplace. The most frequent condition under the "mental health" umbrella is stress and a great deal of effort has been directed to understanding the cause, effects and prevention.

Stress

One observer described "stress" as "a reality like love or electricity – unmistakable in experience but hard to define." Stress is, of course, not confined to the workplace but may be related to home life and the social scene. "Occupational stress" can mean either the pressure that work puts on individuals or the effect of that pressure. All work puts some pressure on individuals; in general, the more demanding the work the greater the stress. This normally leads to higher output and satisfaction with work. However, a point of diminishing returns is reached beyond which increasing stress leads to reversed effects: lowered efficiency, job satisfaction, performance and mental well-being.

   Stress is a normal part of life and the challenge is to manage the pressures so that life is productive and enjoyable. Indeed stress itself is not an illness; rather it is a state. However, it is a very powerful cause of illness. Long-term excessive stress is known to lead to serious health problems. Unfortunately for some, the pressures cannot be managed (the individual may have little in the way of "life skills"), or they may become overwhelming and unrelenting. The assessment of the individual patient requires critical scrutiny of the claim that work stress is responsible for the patient's symptoms; this requires detailed information about the work situation as well as the patient and his or her symptoms. This inevitably leads the doctor to consider the global effects of working practices and the safety of other workers as well as the presenting illness.

Clinicians should recognize that the occupational physician is an invaluable resource in assessing the patient who complains of excessive pressure or stress at work. They will be able to provide the clinician with information about the individual patient's previous work performance and sickness record, and also about the work situation – whether there really are excessive stresses and numerous health problems in that particular department. The occupational physician may also be able to modify the attitude of managers to the patient, negotiate part-time work and observe the patient in the work setting.

The incidence of stress-related illness

Mental illness itself is very prevalent in the working age community. A survey by the Office of Population, Census and Surveys (1994) (Meltzer *et al.* 1994) confirmed that one in seven adults (aged 16–64) living in private households were suffering from some sort of mental health problem, such as anxiety or depression, in the weeks prior to interview. Although it is likely that the prevalence in a working population will be lower, as working populations are known to be made up of a healthier population than non-working groups (the "healthy worker effect"), mental ill-health clearly represents a significant burden to employers. The Health and Safety Executive (HSE) estimated that 80 million working days in a year were lost to employers as a result of mental disorders caused by stress.

In the trailer to the 1990 Labour Force Survey, published by the HSE (HSE 1990), stress or depression caused, or made worse, by work was second only to musculo-skeletal disorders as the most frequently cited cause of occupational ill-health.

Occupational stress, or stress related to work, has been increasingly cited as a cause of morbidity. For workers, stress is often cited as a contributory factor to accidents, job dissatisfaction and illnesses such as coronary heart disease, alcoholism and hypertension. The inevitable conclusion of such data is that mental health problems will affect employees and businesses alike.

The legal impact of stress on employers – statutory and civil

Employers throughout the world are usually required to provide safe systems of work. More recently, employers are being required to assess the risks of work activities and to take steps to minimize the realization of those risks to their employees. Such general requirements of statutory law have been tested in the civil courts. In a well-publicized case (in the UK) of Mr Walker (a social worker) (Croner 1997), Northumberland County Council was found to be in breach of its duty of care. As a result of this case, Mr Walker was awarded damages of £175,000. The true cost of this case to the employer is likely to have been in the region of £500,000, after taking into account legal costs, sick pay, and the cost of ill-health retirement.

In December of 1996, the Disability Discrimination Act 1995 (DDA) became law in the UK. This Act requires employees to make reasonable adjustments to working arrangements where they may cause a substantial disadvantage to the disabled person. Persons with a mental illness which is clinically well-recognized will be regarded as having a mental impairment for the purposes of the DDA. This highlights the importance of obtaining competent advice from an occupational health professional where a person has been or is thought to be suffering from a mental illness. Without such advice, employers may not be in the best position to judge whether they are required to make adjustments. The current focus on mental health risks in the workplace means that this area of disability law is likely to be one of continued legal development.

The financial impact of stress on employers

Estimates in surveys by the United Kingdom's Confederation of British Industry (Sigman 1992) and the Department of Health suggest that up to 30 percent of sick leave in the UK is related to stress, anxiety or depression. The financial costs to employers are thought to be in the region of £3.7 billion each year.

For employers, stress has been cited as contributory to increased sickness absence, and reductions in quality and productivity. Costs are not just limited to those of sickness absence. For many organizations, mental health problems are the commonest reason for early retirement on the grounds of ill-health. This represents a significant cost to the employer, not only in terms of pension benefits but also the costs involved in recruitment, retraining of staff and loss of experienced personnel.

For some employers, such as the emergency services, the hazardous nature of the work is recognized. Nevertheless, certain circumstances may be associated with an increased risk of post-traumatic stress disorder.

In such situations, early provision of appropriate assessment, support and treatment can minimize its impact on employer and employee. Recent awards exceeding £1 million to police officers suffering mental trauma following the Hillsborough football stadium tragedy in 1989, when a large number of people died, highlight the importance of effective prevention.

DESIGNING A FRAMEWORK TO ADDRESS MENTAL HEALTH

Health, particularly the non-physical element, should be addressed comprehensively. The prudent employer will ensure that there is a framework within which those who suffer with mental illness can be helped and referred appropriately. There is also a need to have a comprehensive Model of Well-Being. Figure 9.1 depicts a model which reflects our understanding of the impact of stress on the individual and organization and the elements necessary to combat its effect.

In order to ensure that people feel fulfilled and perform well for the organization for which they work, it is important that they are "healthy" and their well-being is considered. There should be a focus on *work organization and functioning* with proper attention being paid to jobs, to the people who do the jobs in terms of their performance being managed and their motivation being considered, with reward being appropriate. This

Health	Poor (mental) health ———→		Improving health and well-being ———→	Good health and performance
	Serious health problems Depression Anxiety		*Well-being Development and Growth*	*Fulfillment*
Resources	Referral	Advice and Counseling	Health Promotion Training – task – assertiveness – team building – stress/life mgmt	Work Organization Functioning job – structure people – number – selection – development – performance – reward
Agents	GP/Specialist	Occupational Health – doctor – nurse – counselor External Agencies	Occupatonal Health Human Resources Training Line Management	Human Resources Senior Management Line Management

Figure 9.1 Model of well-being ©E. L. Teasdale and R. J. L. Heron (1999)

should be backed up with appropriate *training and education* so that staff are, or become, confident and competent. This should primarily revolve around the tasks and skills required, e.g. assertiveness, team-building, etc. Life-management skills (including stress-management skills) are important and training should be available.

If both *organization* and *training/education* are fully addressed then staff should be "healthy" and be able to perform effectively at their work place. However, many people run into problems from time to time and *advice and support services* (such as counseling) should be available – perhaps by way of an Employee Assistance Program (EAP). Much of this should be proactive to ensure the people have the skills to manage the complexity which is part and parcel of everyday life. Referral will, on occasions, be required.

Stress-management workshops

Zeneca, in the period 1993–1999, ran a number of training and education events for staff – much of this effort being channeled into running workshops. The objectives for these were set as follows:

- to raise awareness of what is meant by "stress"
- to legitimize stress as a subject for discussion in the business
- to show a range of stress-management skills with a view to further skills training
- to practice two key stress-management skills, i.e. listening and relaxation.

The organization and presentation of these workshops was a joint responsibility shared by the Training Section of Personnel/Human Resources and the Occupational Health Departments.

Study – How effective are the workshops?

In 1996 a study was performed to evaluate the effectiveness of these workshops in raising awareness of stress management by assessing understanding of the key principles outlined in a company guideline on the prevention of adverse effects of stress (Zeneca 1996). A secondary aim was to assess the coping skills of attendees, and their self-assessed well-being (Heron *et al.* 1999).

The study was an attempt to measure the input of stress-management workshops on an employee's use of coping strategies, their ability to manage stress in their own staff and their self-rated well-being. Although a prospective study design would be the preferred choice to evaluate the effectiveness of the workshops, a cross-sectional design was chosen to provide more rapid feedback and for the secondary benefit of obtaining information about employees who had not attended the workshops.

The study included 452 UK employees of Zeneca who had attended Stress Management workshops as part of their general development training between 1988 and 1996. They were matched by age, sex and department to employees who had not attended a workshop. A four-part questionnaire was mailed to all participants comprising the following:

1   The 30-question General Health Questionnaire (GHQ30) used to compare self-rated stress/well-being in the two groups (Goldberg 1972).
2   Coping skills questions from the Occupational Stress Indicator (Cooper et al. 1988).
3   Questions assessing understanding of company guidance regarding management of stress in staff.
4   A modified life-events questionnaire (Burns et al. 1967).

The main statistical findings were as follows:

- Mean GHQ30 scores showed no dependence on workshop attendance but there was a strong relationship with the life-events score.
- GHQ30 scores had fallen by approximately 10 percent 2–3 months after the workshop but had returned to their original levels at the time of study (1–8 years after the workshop).
- Female subjects had a better mean coping strategy score than male subjects.
- On average, workshop attendees had only slightly better coping strategy scores than non-attendees, but were three times less likely to be in the group of subjects with the worst coping strategy scores.
- Female subjects, and subjects who managed the most staff, were best able to manage stress in staff.
- Workshop attendees were eight times less likely to be in the group of subjects who were least able to manage stress in staff.

The study failed to show large differences between attendees and non-attendees. A major factor was that the study design was not ideal for the purpose although it did provide a good overall view of levels of stress awareness and self-rated well-being in the business. Another reason may have been that the workshops are only one component of a business strategy to deal with the effects of stress. Consequently it is difficult to isolate the benefit of the workshops, especially in a business where there has been a major change in awareness of stress at all levels, over the time period that the workshops have been running.

COUNSELING AND "LIFE MANAGEMENT"

The "Model of Well-Being" depicted in Figure 9.1 describes the appropriate primary, secondary and tertiary elements which together allow an organization such as AstraZeneca to address well-being and mental health in particular.

1   To ensure success, we focus on organizing work and ensuring good functioning of sections, departments and the total business. Additionally, effort is expended on structuring jobs and developing and rewarding staff. This is largely the responsibility of managers and our Human Resources Department. One key element is our approach to flexible working.

2   To support this, we promote health in its wider sense and try to ensure that individuals are trained for the job (such that they are confident and competent). Many training needs may be apparent, such as assertiveness, negotiating, team-building and, last but not least, skills in life management. Human Resources, Occupational Health staff, and the CALM Team: our training resources and managers are responsible here. Our "CALM" Program was set up in the UK three years ago – CALM covers Counseling And Life Management. We are seeking to ensure that services are available to support our staff when they have difficulties, but also give them skills in life management so that they can lead balanced, healthy lives.

3   On the left-hand side of the "Model" is displayed progressively worsening health. Here we need to ensure that advice and counseling are available and the Occupational Health staff, our doctors, nurses, counselors and external agencies help us in this area. Obviously we work with General (family) Practitioners and specialist staff, e.g. psychiatrists and psychologists.

CALM is an initiative currently under way and aimed at addressing the needs of individuals who are asking for additional support to cope with busy lives (a reactive approach). CALM also aims to enhance the awareness and skills of our staff to enable them to manage their whole lives more effectively for their own personal benefit and, ultimately, for the organization as a whole (a proactive approach). The CALM team includes three trained counselors. They have extensive experience from their work with Zeneca, and other employers, over a period of 10 years, giving them a much greater insight into the culture than might be available to an external support program. And yet they retain a distance and the confidentiality necessary to ensure that staff feel able to approach them with confidence.

In setting up CALM, we also drew on the experiences of the support programs offered by Zeneca Inc. in the USA as well as other models of "best-practice" outside Zeneca. The CALM team has tried to address the needs of employees at all levels through a steering group with representatives of Occupational Health, Human Resources, senior management and site employee representatives. We have also been able to update the most senior members of the company with our plans and progress and hope to keep them aware of the broad issues which the team is addressing whilst, of course, protecting the confidentiality of individuals who seek to use the service. The first six months saw the development of a code of practice and a wide circulation of two brochures, one explaining the benefits of CALM and the other helping employees to develop improved ways of balancing commitments, whether at home or at work. Over 6000 (Zeneca) employees now have access to individual counseling and the early indications are that the service is being well used.

CALM is now looking at the needs of employees distant from its major sites in the northwest of England, as well as putting together programs to address specific issues for groups. With the backing of the business, the steering group, and the undoubted commitment of the team itself, we believe this is a significant initiative for the benefit of employees and the business alike.

Current and future focus

Many companies and organizations have majored on stress-management training and the provision of counseling/EAP programs. Whilst these are valuable elements, skills training should be primarily focused on helping individuals to build mental resilience. Skills to appreciate the complex, fast-changing world and adapt to its rigor are essential. Increasingly, also, a focus on balanced living is seen as critical. Paying attention to all areas of one's life is essential to ensure both sanity and enjoyment.

Summary

Defining objectives and accountability clearly, setting priorities and managing time effectively are essential. In an innovative and demanding environment, maintaining the health of staff and managing stress positively is likely to improve productivity, reduce errors, increase creativity, improve decision-making and lead to enhanced job satisfaction.

A policy for mental health is not a "stand-alone" initiative, but part of an integrated approach to managing a high quality organization. The challenge in any organization is to allow and encourage an appropriate amount of stress or pressure to enhance the performance of individuals, the departments where they work and thus the business and organization as a whole.

The good "health" of the organization is likely to be best served, with effective working, by ensuring employee health and well-being.

References

Burns, L. A. (adapted) from T. E. Holmes and R. H. Rahe (1967) The social readjustment rating scale, *Journal of Psychosomatic Research* 11: 213–18.

Cooper, C. L., Sloan, S. J. and Williams, S. (1988) *Occupational Stress Indicator*, Windsor: NFER-Nelson.

Croner (1997) *Croner's Healthcare and Safety at Work, Special Report.* Issue 28, April 1997, *Stress.*

Goldberg, D. (1972) *The Detection of Psychiatric Illness by Questionnaire*, Oxford: Oxford University Press.

Heron, R. J. L., McKeown, S. P., Tomenson, J. A. and Teasdale, E. L. (1999) Study to evaluate the effectiveness of stress management workshops on response to general and occupational measures of stress, *Occupational Medicine* 49(7): 451–7.

HSE (1990) *Trailer to the 1990 Labour Force Survey.*

Meltzer, H., Gill, B. and Pettigrew, M. (1994) *OPCS Survey of Psychiatric Morbidity*, OPCS.

Sigman (1992) *Report by Confederation of British Industry.*

Zeneca (1996) *Prevention of adverse effects of stress*, Zeneca SHE Guidance Note 9.6.

Chapter 10

Designing and implementing a managing pressure program at Marks & Spencer Plc

S. Williams and N. McElearney

THE BACKGROUND

Marks & Spencer Plc has, for many years, been described as Britain's leading retailer. The business is UK based with a significant and growing international presence and currently trades in 683 locations worldwide. In the United States it owns 191 Brooks Brothers stores and 21 Kings Supermarkets. It has about 60,000 employees, the majority of whom work in the UK. Of the 45,000 who work in the UK-based company, about 85 percent are female. The majority of staff work part time.

Marks & Spencer has an enviable reputation for being an excellent employer. Its staff are extremely loyal to the company and take an enormous amount of pride in working for Marks & Spencer. When asked what they do for a living, Marks & Spencer staff will not describe themselves as sales assistants or working in the retail sector or store managers; their immediate response is to say they work for Marks & Spencer. A strong factor in the business's reputation is the health and welfare provision it provides for its employees. Marks & Spencer was one of the first companies in the UK to offer its employees health screening. It has provided cervical and mammographic breast screening since the 1970s and it also provides cardiovascular risk factor screening for all its staff on a voluntary basis.

Health services department

The company's Occupational Health Service is part of the Personnel group, and has a centralized structure that provides a clinical Occupational Health Service and manages the various programs. The service is led by a team of occupational physicians based at the London head office and supported by store-based sessional Occupational Physicians, full-time Occupational Health Nurses, and administrative staff. The Health Services Team is responsible for setting the corporate healthcare policy and designing and delivering programs to meet their objectives. Marks & Spencer has had healthcare provision since the 1930s. At each stage in the development of the now complex offer, the team recognized gaps in state provision and persuaded the company to provide additional healthcare for its staff. Although it would not have been stated in such terms, value has been added by offering services that the staff would otherwise not have been able to get by themselves. For example, cervical pre-cancer screening was offered to all female staff several years before the national program commenced. Mammographic breast screening is offered to all women aged 40 and above, every two years, with bilateral views each time. Trials are under way to examine the efficacy of oral pre-cancer screening. The managing pressure program is another example of Marks & Spencer tailoring an innovative health program to its exact requirements.

Identifying the need

Over the past few years the retail sector has become increasingly competitive. There has been pressure to extend opening hours, reduce overheads, adopt new technology and continually improve customer service levels. In the early 1990s the Health Services Team and Store Operations Management recognized that the demands on staff were increasing and that, although not currently a problem, stress was starting to become an issue. At that stage the problem was seen as one for general staff. The company commissioned Resource Systems, an independent consultancy specializing in managing pressure at work, to investigate the problem and design appropriate interventions. Resource Systems worked with the Health Services Team, store managers and staff at all levels to identify the training need and produce a proposal for addressing this need through a series of targeted interventions.

The needs analysis showed that staff felt that they were working under increasing levels of pressure and that with the continuing increase in competitiveness of the business this would get worse not better. As part of its existing healthcare provision the company had excellent programs for treating people suffering from stress-related illnesses. Staff could be referred to their own family physicians or directly to psychologists. However, little was being done at a primary or secondary level to help staff manage pressure more effectively so that stress-related illness could be avoided. There was an anxiety that raising awareness of the issue could create a problem that could not be controlled.

The project team

In order to develop and implement the managing pressure program, Marks & Spencer established a small project team with representatives from Occupational Health Services, Training and Development, and the external consultants. The representative from Training and Development was there to ensure that the program met the needs of the business and complied with the best practice in training in Marks & Spencer. The Occupational Health Services representative ensured that the program accurately met the needs of employees and addressed the issues of stress and pressure that ultimately manifest themselves in physical or psychological ill health. The external consultants ensured that the program was built on a detailed understanding of stress at work, and the most effective methods of improving the ability to manage pressure.

The objective

It was decided to develop a training program for the area of greatest needs, i.e. staff in the stores. Although the business recognized that there were major pressures at head office and amongst the management teams, it was felt that the customer-facing staff – the first-line employees – were the ones that would benefit the most from the program. In making this decision the project team recognized that conventional wisdom would be to train the managers first and then cascade this training down to the staff. However, in order to provide immediate benefit to the staff with the greatest need, and to help to demonstrate the value of the training, the first programs were directed towards supervisors and sales assistants in stores.

Management involvement and "buy in" to the programs was obtained through a cascade process in which the divisional training managers and divisional personnel managers were briefed by the project team and they, in turn, presented the program outline to their personnel and training managers. The message was then passed through to the store management by their local personnel and training managers. This communication process was supported by the occupational health advisors talking to store management and staff in more detail about the objectives and content of the program. Information flows in all directions in the organization and a key part of the briefing process was to open a conduit so that information from the stores could flow back to the project team and help to refine the program design.

The development process

Marks & Spencer has a long tradition of continuously improving its products and services. It continues to reformulate and redesign even its most successful products and every aspect of the business is involved in a constant search for a better way of doing things. The program development was based on the premise that a series of workshops would be developed, tested, modified, retested, redeveloped, and so on until the company was satisfied that the workshop met the needs of the staff and the business and could then be rolled out on a wider basis.

The design philosophy

The workshops and supporting materials were designed on the basis that emphasis should be given to the positive management of pressure, not the remedial treatment of stress. It was also considered important that the business adopted a consistent and systematic approach to stress at work and that benefit could be gained by coordinating various local initiatives to ensure consistency of approach. This was summarized by the occupational physician as "pick a definition and stick to it." The program had to be accessible and available to staff at all levels, and therefore the material and course content should be easy to understand, irrespective of educational background or other factors. The training needs analysis had shown that stress can be caused by home and by work and that although this is a work-based program the impact of the home cannot be ignored. In any event, whether the stress-related ill-health comes from home or work is irrelevant. The impact on the bottom line is the same.

Basic principles

The program was developed from three basic principles:

- Pressure is inevitable in working life and an individual can actively manage their response to pressure, to improve the probability that pressure produces positive not negative outcomes.
- The managing pressure program recognizes that there is a balance of responsibility between individuals and organizations. Sometimes managerial or organizational interventions may be the only way of alleviating the sources of pressure.
- There is a direct relationship between pressure and performance. Improving an individual's ability to manage pressure improves their performance. The program is designed to be positive and developmental. It is of direct benefit to both the individual and the company.

The intervention model

Our experience suggests that most people find awareness of the stress process to be a major factor in helping them to manage pressure. People may have difficulty in seeing the extent to which their behavior changes when they are under stress and fail to recognize the effect stress has on their physical and mental health. The managing pressure program was designed to raise awareness of the issues, show the delegates that they have a choice, encourage them to take responsibility and motivate them to take action in the most appropriate and comfortable manner.

The workshops were therefore structured to:

- raise awareness
- give information
- share experiences
- encourage self-discovery.

The intervention model is based on the premise that simply providing information about pressure and stress is unlikely to lead to sustainable improvements in an individual's ability to manage pressure. The readiness to change models developed by Prochaska and DiClimente (1983) show that individuals will go through various stages before lasting change is achieved. Prochaska and DiClimente's five-stage model identifies these stages as: Precontemplation, Contemplation, Preparation, Action, and Maintenance. People need different messages depending on where they are in the change cycle and it is extremely unlikely that every delegate on a course will be at the same stage or have the same need to change. Raising awareness, helping individuals to understand their personal stress profile, and providing relevant information all help people to help themselves to find their own solutions.

The OSI/PMI questionnaire

It was felt that the course would be much more relevant and, therefore, much more successful if each delegate received a personal stress profile showing the effects of stress on them, their sources of pressure, individual behavioral differences and coping skills. The personal profile was produced from the Occupational Stress Indicator (OSI) (Cooper *et al.* 1988), later replaced with the Pressure Management Indicator (PMI) (Williams and Cooper 1996).

The PMI comprises eight integrated questionnaires designed to identify stress-related problems in organizations. It includes a biographical questionnaire; measures job and organizational satisfaction, organizational security and commitment, and current mental and physical health; it contains an assessment of behavior patterns; a locus of control indicator; a review of sources of pressure at work and an assessment of coping skills. The PMI is based on an extensive body of occupational stress research and is designed to identify and to cross-reference four key elements; the Sources of Pressure, Behavioral Characteristics, Coping Strategies and the Effects of both individual and organizational stress. It has been designed to provide a "working model" of stress that facilitates the production of a practical action plan for change and improvement. The PMI is a self-administered questionnaire that generally takes about 15 to 20 minutes to complete.

This questionnaire was completed by each delegate one week before the course and a computer-generated detailed narrative report for each individual was given out during the workshop. As well as helping each delegate to focus on their specific needs the results from the delegates were aggregated to give an indication of the key issues facing the workshop group. This information helped the trainers to prioritize the various learning modules and adapt each workshop to the specific needs of the group. The other benefit of using the questionnaire was that it encouraged any individual with concerns about their well-being to arrange a confidential one-to-one discussion with the health advisor. Figure 10.1 shows the scales of the PMI.

Positioning the program

A key factor in the ultimate success of the program was to position it from the beginning as a positive personal development workshop for successful people. There is still a stigma associated with stress at work and, in many organizations, going on a "stress management" course is a sign of weakness or failure. Comments like "Only people who can't cope go for stress training" reflect the "stress is for wimps" attitude found in many organizations. At the time the programs were introduced stress was still an extremely sensitive issue in the UK and a great deal of care was taken to ensure that people were not labeled as being stressed or seen to

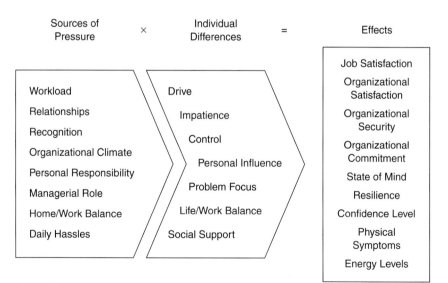

Figure 10.1   PMI Scales

Source: Williams and Cooper, 1999

be failing if they attended the program. Indeed this principal was so important that the title of the workshop and the words used in describing it were chosen to avoid negative labels and emphasize that this is a course for everyone. This theme was continued in the course material and even in the way it was discussed in the business. The need to emphasize the positives was also a major factor in the choice of stores used to pilot the workshops. These stores were chosen on the basis that they were amongst the best in the business, both in terms of their commercial success and also their management of people. Careful attention to the "branding" of the program and the explicit focus on the management of pressure not the treatment of stress was a key factor in gaining acceptance throughout the business. The programs have continued to be seen as developmental not remedial, and attendance enhances rather than damages opportunities for career progression.

THE STORE-BASED MANAGE YOUR PRESSURE PROGRAM

Version I – The two-day program

Program structure

The first workshop was designed as a two-day event that would be run by an occupational health advisor (OHA) and the personnel manager. It attempted to include almost everything anyone would need to know about stress and how to cope with it. The program was built around a series of modules which take into account the constraints on the business of running training programs within the store, in particular the effect on staffing and rotas. The basic component of the program was a comprehensive delegate workbook containing over 100 pages of notes, activities and pre-course work. Each section of the workbook was integrated with a series of interactive workshops. The workbook and workshops are linked to pre- and post-course diagnostic instruments and supported by a comprehensive range of learning materials, training courses and individual advisors. Figure 10.2 shows the main elements of the program and the way it links to existing Marks & Spencer services.

In order to minimize the time that people were off the sales floor, the workbook was intended to prepare the delegates for the workshops by giving them a background understanding of the subject and presenting them with a range of exercises to be completed before the workshop sessions. Even though staff were given time to complete the pre-course work during working hours, very few of them managed to do any preparation before the course. Staff found the workbook interesting but found it difficult to get involved in the material.

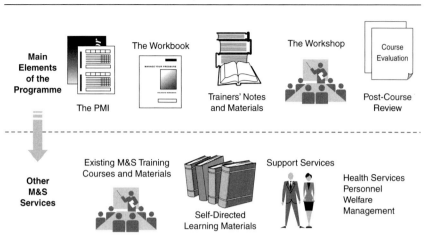

Figure 10.2 Program components

The agenda was as follows:

- Introduction
- Exercise 1 – Who are we? Delegate introduction and personal objectives
- Exercise 2 – The effects of stress
- The signs and symptoms of stress
- Exercise 3 – Sources of pressure
- Pressure and performance
- Managing pressure model
- Exercise 4 – Positive choices
- Appropriate interventions
- Exercise 5 – Appropriate interventions
- Action planning
- Review and close

This program was tested on 4 groups of 12 employees based in 2 different stores. One of the most valuable elements of the program was the exercise to identify specific sources of pressure. In this exercise the delegates worked in two groups of six to identify and subsequently prioritize their sources of pressure. Table 10.1 gives an example of the sources of pressure reported by the supervisors, and Table 10.2 the pressures reported by sales assistants.

Listing major sources of pressure enables common themes to be identified, helps people to see that "they are not alone" and provides further information about organizational and managerial issues that can be amalgamated and fed back to the business without breaching confidentiality.

Table 10.1 Supervisors' pressure

No time available from management	No support
Instant reaction	The actual job
Changes in management	Working to budgets
Absences	Time limitations
Lack of communication between individuals	Regionalization
Staff restructure	Inexperienced staff
Feeling isolated	Everything too serious

Table 10.2 Sales assistants' pressure

Having too many jobs at once	Working "six" days every week
Work deadlines, jobs that *have* to be done	Christmas
Getting till queues down	Lack of communication
Lack of staff	Lack of equipment
Getting up *very* early	Customers
Getting into a routine	Late breaks, not getting a break
Being asked to do "extra" hours	Getting orders in "on time"
Heavy workload	Not enough time
No praise or recognition	Long hours
Favoritism	Too many responsibilities
Ignorant management	Lack of support

The "Pressure list" also proved to be a very effective mechanism for helping the delegates to understand the benefits of a wide range of coping skills and prompted a discussion on situational coping.

Developing coping skills

Once the main sources of pressure had been identified the workshop moved on to a discussion of various ways of coping with pressure and an exercise to decide which would be of most benefit to the delegates. Table 10.3 illustrates the ranking system used to decide which coping mechanisms would be addressed in the time available. The exercise involved giving each delegate three different colored dots, one color for their first choice, one for the second and another for their third. They topics were displayed on a flip chart and the delegates were asked to put a dot against the topic they would be most interested in learning more about.

This exercise was very popular and helped to keep the workshop relevant to the needs of the group. An additional benefit of this "let the delegates decide what they want to learn" approach was that the summary results for each of the pilot groups could be used to determine which topics were included in the next version of the program.

Table 10.3   Coping mechanism ranking system

Positive choices – top four coping mechanisms

Topic	Sales assistants' choice			
	1st	2nd	3rd	4th
Leaving stress at work	6	1	1	0
Time management	1	1	0	1
Communication skills	0	0	3	2
Social support	0	0	0	0
The home/work interface	0	3	0	1
Taking control	0	0	0	0
Balance	2	0	2	1
Positive thinking	0	3	2	3
Humor	0	0	0	0
Diet	0	0	0	0
Physical fitness	0	0	0	0
Goals	0	0	0	0
Relaxation	0	1	1	0

Version 2 – The second pilot program

Staff found the initial pilot program highly beneficial; however, it was felt to be "comprehensive;" in other words it was too long and too detailed! It was decided to shorten the workshop to one day and simplify the course materials. Staff were very clear about the bits they thought should be trimmed from the course and what should be left in. It was recognized, for example, that asking staff to do the pre-course work was unrealistic and this was removed from the program. However, completing the diagnostic questionnaire and receiving a personal pressure management profile was regarded as a great benefit and an essential part of the workshop.

Version 3 – The half-day program

The evaluation of the second program showed that it worked well and the training materials were well received. However, it was felt that having a member of the personnel team involved in delivering the program was detrimental to the levels of confidentiality and honesty necessary to encourage full and open participation from the delegates. This wasn't a criticism of the individual personnel managers but reflected the perceptions of their role in the business, in particular their staff management responsibilities. In contrast, the occupational health advisers were well respected within the business and clearly associated with medical confidentiality. Staff felt that they could talk openly to the occupational health advisers without fear that their comments would end up on their personnel files or be fed back to store management. Taking the personnel managers away from the program increased the perception of confidentiality.

The other factor in the decision to shorten the program was the feedback from the store managers on the practical implications of releasing staff from the sales floor for a full day. Although some management teams would be willing to release some staff for a full day, the only way the workshops could be made available to large numbers of employees in all stores was by reducing them to half a day.

The final content of the half-day workshop was based on detailed feedback from delegates and all of the trainers. An additional benefit of having included personnel managers on the earlier versions of the program was that they were able to offer their advice on the content and were also more supportive of the new format.

The final version of the Manage Your Pressure workshop takes four hours to run and takes place on one day, usually from 9:30 am to 1:30 pm. The agenda is shown in Figure 10.3.

Training the trainer

Health Services selected 12 occupational health advisers to deliver the original courses. All had been trained in working with groups and had good interpersonal skills. About another 20 have been added to their number since then. Their preparation for the workshop has remained

---

Part One: Understanding

1. Introducing the program
2. How can I recognize stress?
3. What is the difference between pressure and stress?
4. Why does pressure sometimes lead to stress and sometimes to better performance?
5. Where does pressure come from?
6. What does my PMI profile mean?

Break

Part Two: Action

7. Taking control
8. How do I manage pressure? Tips for short-term coping
9. How do I manage pressure? Getting better at bouncing back
10. How can I prevent stress?
11. What should I do next?

Figure 10.3   Manage Your Pressure workshop agenda

unchanged. Each occupational health adviser attends a one-day "Train the Trainer" course before being allowed to run the workshops. They receive a trainer's pack containing all the materials they need to run the workshop. The content of the Train the Trainers program has developed in parallel with the various versions of the workshop and feedback from the trainers has been an essential part of the continuous improvement process.

The trainer's pack for the current workshop comprises:

*   the *Manage Your Pressure* booklet
*   the trainer's guide
*   overhead transparencies (OHTs) to illustrate key points
*   draft delegate's briefing letter
*   delegate packs:
    – delegate's workbooks
    – stress dots
    – PMI questionnaires
    – personal action plan leaflets

The trainer's guide builds on the information in the *Manage Your Pressure* booklet. It includes a timetable for pre-work, a structure for the workshop, suggestions as to how each session should be run, copy OHTs, and a copy of all the workshop activities as set out in the delegate's workbook.

The delegate's workbook contains the working papers for the activities that delegates will work on during the workshop. Delegates are told that it is not a self-study booklet and will only give value if used during and after a workshop session.

Rolling out the workshops

The success of the store-based "Manage Your Pressure Program" confirmed the need to deliver the basic messages to as many people as possible in a cost-effective way. It was decided to adopt a three-tier strategy for achieving the widest and most relevant exposure to the key concepts of managing pressure. Constraints on training resources, staff availability and funding meant that it was not feasible to put over 35,000 people through the half-day workshop in a reasonable time scale. It was also recognized that the managers would benefit from an extended program that would include sessions on management style and a more detailed understanding of the influence of management behaviors on other people's pressure.

The approach was therefore to:

1   Produce a booklet to be used by all staff based on the developed material.

2   Increase the number of trainers and make the workshops available on request.
3   Develop a "Managing Pressure for Managers" program.

Manage Your Pressure booklet

The fastest and most cost-effective way of delivering the core principles of the Manage Your Pressure workshops to a large number of employees was via a booklet. The training materials were further developed and condensed into a *Manage Your Pressure* booklet. The initial drafts of this booklet were written by the consultant and when the final contents had been agreed the entire booklet was rewritten by a journalist to simplify the material and ensure that the reading level was equivalent to that of a daily tabloid newspaper. Twenty-thousand copies of this booklet were produced in 1997 and the vast majority were taken by staff within a couple of weeks of release. The booklet proved so popular that a further 15,000 copies had to be printed a few weeks later.

The booklet was supported by a poster campaign designed to raise the awareness of managing pressure at work and to make it more acceptable for staff to discuss pressure and stress with their supervisors and the management team. The booklet and poster campaign proved extremely successful in raising awareness of pressure as an issue and encouraging staff to focus on the positive aspects of managing pressure rather than seeing stress as a weakness. Staff welcomed the booklet and the concern that the business showed about their mental well-being. Health services continue to receive regular feedback from staff saying that they value the investment in them, and that the booklet is beneficial for helping them to manage home- as well as work-related pressures.

The *Manage Your Pressure* booklet talks about the difference between pressure and stress. It looks at where pressure comes from and how to manage pressure so that it doesn't lead to the harmful effect of stress. It gives tips for the short term but also looks at building long-term resilience. Finally, it shows how to make a personal action plan and how to act on it. The booklet can be used for self-study as well as being a key document in the Manage Your Pressure workshop.

Managing Pressure for Managers program

The success of the store trials and the *Manage Your Pressure* booklet created an environment in which staff and the management teams were asking for a program suitable for management. The project team thought that this should follow the same themes as the staff-based program but should go into the subject in more depth and place greater emphasis on the manager's responsibility for helping their staff to reduce pressure.

The program was intended to provide immediate benefit to the managers attending the program by giving them tools and techniques they could use to manage their own pressure and avoid stress. It was also intended to help them to analyze stress at work, recognize the issues and, in the best principals of primary prevention, remove pressure at source. As with the other programs the development cycle for the Managing Pressure for Managers program went through a number of stages.

The first trial

The original manager's program was designed as a one-day workshop that would help groups of managers to understand pressure at work and take action to manage it more effectively. It was felt that the program would be more effective if it dealt with the whole area of individual and managerial well-being and was therefore built around physical as well as mental health. The interlocking themes of home, health and work formed the basis of the program material and the workshops were delivered by an occupational physician, an organizational psychologist, and an exercise physiologist.

It very quickly became apparent that it was not possible to cover the material in any depth in such a compressed time-scale. The delegates' evaluation was that the workshop needed to be at least a day and a half, that it should be residential and more experiential. The Version 2 work-

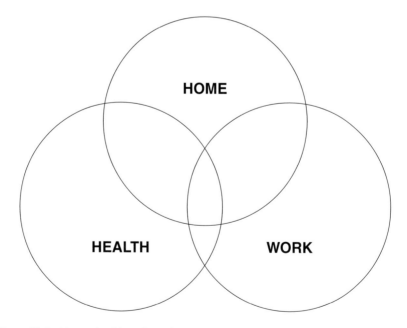

Figure 10.4   Home, health and work

shops were therefore extended to one and a half days and trialed with various management groups from different parts of the business. The content and delivery was fine-tuned after each workshop in the series. For example, an early morning run was included to reinforce the physiology messages. Each workshop gave more insight into the demands facing the manager groups and the ways that people manage things for themselves. The feedback confirmed that the course was very good at giving people relevant information that was accessible but without the certainty that it would lead to action. The course content for this program is shown in Table 10.4.

Table 10.4   Course content

Session	Content
The organizational issues	Why are we trying to tackle the issues? What is the strategy?
About PMIs	What do they mean? How to interpret your own Interpreting your PMI profile
Some basic exercise physiology	
Exercise session What is stress?	The physiology of stress Stress is a normal physiological event Fear, fight or flight, soldier's heart, metabolism The psychology of stress Symptoms and signs Recognizing stress
Artificial adrenaline So where does pressure come from?	Home, work and health – the "top five lines" Sources of pressure The relationship between pressure and performance
Diet Solutions	Taking control The need to develop resilience/bounce back Matching coping techniques Solutions to the "top five lines"
Diet, exercise and relaxation	Breathing Coping techniques in the workplace Meetings Volume/detail/delegation/prioritization Asking for help Saying no Managing pressure in yourself Managing pressure in those who work for you Developing commitment and attitude Relaxation

From managing pressure for managers to maximizing your potential

The latest phase in the development of the managers' program has been a change in emphasis away from giving information towards working through a process that increases each individual's ability to change. The program now makes extensive use of the principles of Solution Focused Brief Therapy as developed by Steve De Shazer and his associates in Milwaukee. The workshop concentrates on recognizing and utilizing the expertise that people have for fixing their own problems. The group work helps delegates to realize that, for some of the time at least, they are on top of the issues and know what to do and when to do it. Although the workshops continue to take a holistic approach to health and well-being and provide information on home, health and work issues, the majority of the time is given to developing individual solutions to problems.

The program is residential and runs over an intensive 24-hour period starting at 4:00 pm on the first day with the formal program, and stopping with dinner in the evening. The second day starts at 7:00 am with a pre-breakfast exercise session including a 20-minute run. Figure 10.5 shows the agenda and the objectives of each session.

This workshop has been extremely well received by the managers and has forced them to give detailed and careful thought to the relationships between home, work, and health and produce specific sustainable solutions to the major problems facing them in these areas. For example:

- "to have open, honest, motivating relationships within management team"
- "to be involved and contribute towards decision making"
- "to work in an honest way which I can influence"
- "work in an environment where achievements are recognized and I feel valued"
- "achieve personal satisfaction at work."

Identifying the issues

Early in the course the delegates were asked to choose five things from each of home life, work life and their personal health that they would most like to change and rate the importance of each of these issues on a scale from 0 to 10. This is followed by an exercise on transforming these problems into corresponding goals, which in turn sets the agenda for small group work. The delegates then work with these goals to produce detailed sustainable improvements on the issues that matter. As in the store-based programs the Pressure Management Indicator is used to provide independent individual feedback to the delegates on their personal stress profile.

The results for the group are aggregated to show the extent to which they differ from other management groups. Table 10.5 shows the top five key issues for one group of managers in a part of the business going through major change. The percentage scores for these items is a measure of the

---

INTRODUCTION AND INDIVIDUAL OBJECTIVES

The solution focused approach, deciding what needs to be fixed. Working with what you've got, understanding your body and mind. The process of transforming problems into goals, working through some examples.

By the end of this first session you should be familiar with your sources of pressure in general, understand that stress is an outcome. You should be beginning to think about ways of finding your own solutions in relation to your health and home issues.

RUN FOR YOU LIFE

The learning objective is to put you in touch with your physical response to effort. The benefits of exercise are also explained. You will need suitable clothing and footwear as most venues have outdoor facilities for a short jog. Talk to a tutor if you have a health issue or any concern about this.

FROM GOALS TO SOLUTIONS

The process of incremental change, finding exceptions and using scales. Getting your body to help. Understanding the links between body and mind.

The learning objectives in this session are to introduce you to areas where you have already have a high degree of control and to give you an opportunity to practice a solution focused approach to these issues.

CAFFEINE – trouble in a cup!

We *will explain how to use this artificial adrenaline to your advantage.*

FROM GOALS TO SOLUTIONS AT WORK

The top five problems transformed into goals. Looking for examples of changes that have already occurred. The issue of trust. Incremental improvement, a follow-up process.

This is your opportunity to tackle the big issues of the day. The process will demonstrate that even difficult issues present opportunities to create change. We will work through a couple of examples that are specific to your group to ensure you understand the approach.

DIET – you are what you eat

Lunch is specially prepared to allow you to choose what you want to eat in a healthy way. The session should give you a framework for making those decisions.

FROM WORDS TO ACTION

Type A personality, taking positive control, techniques for increasing your personal resource pack. We will revisit exercise and have a relaxation session. We will also cover motivation and commitment to change.

This final session is focused on practical activity so that you can put into operation what you have learned.

Figure 10.5 Managing pressure for managers' agenda

Table 10.5   Top five key issues

Issue	Percent difference from other managers
Characteristics of the organization's structure and design	26
An absence of any potential career advancement	24
Lack of consultation and communication	24
Morale and organizational climate	22
Demands their work makes on family relationships	21

extent to which their average score differs from a comparison group of managers in the private sector.

This list of issues, calculated from the items on the PMI Sources of Pressure scale, is used to focus attention on the major work-related problems facing the group. In this particular example, the top five issues were very different from the issues identified by the analysis of the PMI items for other groups working in other parts of the business. The use of the PMI to identify relevant issues for each group, combined with the flexibility of the workshop and the adaptability of the process, means that every workshop is relevant to the needs of that group. The workshops deal with real, specific issues that are important to each group of delegates. This adaptive, diagnostic-based approach produces specific solutions to real issues and is far more relevant to the needs of the delegates than prescriptive, content-based training. The independent perspective provided by the PMI, together with the delegates' assessment of key workplace issues, creates a very powerful impetus to change. The technique goes on to examine where elements of their preferred future are already happening and provides a practical route to incremental improvement that is achievable and sustainable.

Measuring the effectiveness of the workshops

Changes in pre- and post-course PMI results

One of the most difficult tasks in the design, development and implementation of training programs is the evaluation and measurement of effectiveness. The process for formally evaluating the effectiveness of the workshops by a pre- and post-assessment has been fraught with difficulties and illustrates the problems of conducting research in a dynamic business environment. The original concept was to carry out a pre- and post-course evaluation using the PMI as a measurement tool. The intention was to use the questionnaires completed as part of the pre-course diagnostics as a time one measure and ask the delegates to repeat the questionnaire six months later to give a time two measure. To control for changes in the business a number of employees, selected to match the demographics of the delegates, were asked to complete the PMI questionnaire at time one and again at time two. The rate of change in the business made it obvious that a simple pre- and post-course comparison would be of little value. Unfortunately it was extremely difficult to get the staff in the control group to complete the time one questionnaire and almost impossible to get them to complete the follow-up. Only six people in the control group actually completed both questionnaires, too few to provide a meaningful comparison. It is intended that the lessons learned from this experience will be used to design and implement a more formal pre- and post-course assessment of the latest version of the programs.

Qualitative evaluation

Marks & Spencer is a very pragmatic, results-oriented business. If things don't work they are changed or discarded. In the current economic climate the business will invest in what is seen to work and the ultimate measure of success for any intervention program is the extent to which it adds value to the business. The continued support for the various Managing Pressure programs and the increased investment in time and resources is a strong indicator of their perceived value. The evidence for the success of the programs comes from a wide variety of sources. Course evaluation forms are a standard way of assessing the benefits of training programs, and Table 10.6 shows the average rating for three of the managers' programs. The evaluation forms were sent out to the delegates approximately two weeks after the course in an attempt to measure the value of the training after the euphoria of being on an enjoyable course has worn off.

Table 10.6 Evaluations

	Home	Work	Health
Overall course			
Usefulness rating	8.0	8.15	8.25
Intention to use what has been learnt	100%	100%	95%
Have already started to use what has been learnt	90%	95%	100%
Ability to understand the solution focused approach	8.65		
Confidence that course meets specific objectives			
Awareness of stress	9.25		
Ability to manage yourself	8.85		
Ability to manage stress at work	8.05		

The following comments are typical of the reaction to the training:

1 "My score reflects my ability to make the necessary changes/disciplines, not your ability to prepare me."
2 "Please can we have a follow-up session after we have experience using it?"
3 "I have used various techniques at home and have taken 'baby steps' to improve a stressful situation at home."
4 "An excellent course looking at the individual."
5 "Probably the most relevant course I've attended."
6 "Excellent course – best one I've ever attended in Marks & Spencer."
7 "This has undoubtedly been the best two days I have spent on a course/seminar for a great many years. The messages, presentations, exercises and group activities have had a profound influence on my thinking."
8 "I am convinced that I am better equipped as a manager and team player."

Feedback: All scaled 1–10, where 10 is the most favorable outcome. The individual feedback confirms the willingness of the delegates to enthusiastically endorse the benefits of the program and encourage their colleagues to request further events.

The future of the programs

Marks & Spencer is renowned for its attention to detail and for its desire to measure and control. In the final analysis the effectiveness of the program depends on whether or not it is seen to add value to the individual employee and to the business. Although difficult to quantify, the success of the program is unequivocal.

Over 95 percent of delegates have started to use the technique and information at work and at home. They are not waiting for the program to reach their subordinates, they are showing them how to use this new approach. A similar percentage thinks the course should be part of core management development and believe it will add to the competitive advantage of the company. What could be more attractive for an organization than finding out what it is doing well and doing more of it?

References

Cooper, C. L., Sloan, S. J. and Williams, S. (1988) *The Occupational Stress Indicator*, Windsor: NFER Nelson.
Prochaska, J. O. and DiClimente, C. C. (1983) Stages and processes of self-change in smoking: towards an integrative model of change, *Journal of Consulting and Clinical Psychology* 51: 390–5.

Williams, S. and Cooper, C. L. (1996) *The Pressure Management Indicator*, Harrogate: RAD Ltd.

Williams, S. and Cooper, L. (1999) *Dangerous Waters: Strategies for Improving Wellbeing at Work*, Chichester: John Wiley.

Chapter 11

Developing healthy corporate cultures by reducing stressors at work

C. L. Cooper and L. R. Murphy

The chapters in this book presented a range of different approaches for creating healthy work organizations. Some began with an empirical assessment of the work environment and organizational function and then statistically identified factors that were correlated with employee well-being and organizational effectiveness. Others put in place company-wide programs that addressed a broad ranged of employee health issues.

The thesis of this chapter is that efforts to reduce worker stress will lead to the creation of healthier and more productive work environments. The logic behind the thesis is straightforward. Some of the most common work stress problems identified in the research literature over the past 20 years coincide with characteristics of healthy work organizations proposed in chapters of this book. For example, lack of participation in decision-making (or worker involvement) has been cited often in the stress research (Cooper and Marshall 1976; Holt 1982) and likewise appears as a key element of healthy work organizations. Other examples are pay/performance rewards, open communication, opportunities for career development, supervisor support, and worker autonomy/control.

This overlap of key concepts suggests that efforts to reduce sources of stress at work, if successful, represent a viable strategy for creating healthy work organizations. Since a great deal more research has been done on job stress and stress interventions than on healthy work organizations, it is useful to review the research evidence and summarize the state of the art.

LEVELS OF INTERVENTIONS

There are three basic levels of interventions which have been used to address stress at work. These are termed *primary* (e.g. job redesign), *secondary* (e.g. stress management), and *tertiary* (e.g. workplace counseling), and each addresses different stages in the stress process (Murphy 1988). Each of these approaches will be described in detail and evidence pertaining to their effectiveness summarized.

Primary prevention

Primary prevention is concerned with taking action to modify or eliminate sources of stress in the work environment, and thus reduce their negative impact on the individual. The "interactionist" approach to health promotion (Cox 1978; Edwards and Cooper 1990) depicts stress as the consequence of the "lack of fit" between the needs and demands of the individual and his/her environment. The focus of primary interventions is in adapting the environment to "fit" the individual.

Elkin and Rosch (1990) list 10 strategies to reduce workplace problems:

- redesign the task
- redesign the work environment
- establish flexible work schedules
- encourage participative management
- include the employee in career development
- analyze work roles and establish goals
- provide social support and feedback
- build cohesive teams
- establish fair employment policies
- share the rewards.

The US National Institute for Occupational Safety and Health (NIOSH) advanced another set of recommendations for reducing job stress (Sauter *et al.* 1990). The recommendations, which are not dissimilar from the prior list, include the following:

- *Workload and work pace.* Job demands should be commensurate with the capabilities and resources of workers, avoiding underload as well as overload.
- *Work schedule.* Work schedules should be compatible with demands and responsibilities outside the job.
- *Job future.* Ambiguity should be avoided in promotion potential, career development, and job security.
- *Social environment.* Jobs should provide opportunities for personal interaction, both for purposes of emotional support and for help in accomplishing assigned tasks.
- *Job content.* Job tasks should be designed to have meaning and provide stimulation and an opportunity to use skills.

Unfortunately, studies evaluating the benefits of primary interventions on employee health and well-being are not common in the research literature, and those that have been performed have produced mixed results. For example, early and often-cited reports (Jackson 1983; Quick 1979;

Wall and Clegg 1981) found positive effects of organizational change interventions on worker distress. More recent studies, however, have failed to demonstrate significant improvements in worker distress following organizational interventions (e.g. Heaney *et al.* 1993; Schaubroeck *et al.* 1993). Moreover, recent reviews of the organizational stress intervention literature concluded that such interventions were not particularly effective (Parkes and Sparkes 1998).

This is not to say that companies should not attempt primary stress interventions. The mixed findings from the research literature may simply reflect the fact that organizational interventions are far more complex to design and evaluate than stress management or health promotion programs. For example, it is almost impossible to perform a true experiment in a work setting with random assignment of workers to conditions and control of extraneous factors.

Organizations are dynamic entities and concurrent (and uncontrollable) changes often occur alongside the planned interventions. During post-intervention evaluation, it is difficult to untangle the respective role of the planned intervention and other changes that may have taken place.

A few other reasons for the mixed results can be offered. First, since a fundamental tenet of stress is that change of any type is stressful, then interventions focused on job redesign or organizational change may likely increase worker stress in the short term. This would argue for longer evaluation periods to accurately assess benefits. Second, the change may decrease stress for some workers (the ones who see the change as positive) but increase stress for other workers (those who see the change as negative). This suggests the need for evaluation protocols that include subgroup analyses or that attempt to identify workers who were positively and negatively affected by the interventions.

A final possible reason was suggested recently by Payne *et al.* (1999). Using a large sample of healthcare workers, the authors demonstrated that psychological strain moderated the relationship between work characteristics and work attitudes (i.e. job satisfaction). That is, the relationship between work characteristics and work attitudes was lower for high strain than low strain workers. The authors suggested that stress interventions first should help workers deal with their high strain (through individual-oriented strategies) before attempting to change the stresses in the work environment.

Secondary prevention

Secondary prevention is concerned with the prompt detection and management of experienced stress by increasing awareness and improving the management skills of the individual through training and educational activities. Individual factors can alter or modify the way employees exposed to workplace stressors perceive and react to this environment. Each individual has their own personal stress and health threshold, which is why some people thrive in a certain setting and others suffer. This threshold will vary between individuals and across different situations and life stages. Some key factors or "moderator" variables that influence an individual's vulnerability to stress include their personality, their coping strategies, age, gender, attitudes, training, past experiences, and the degree of social support available from family, friends and work colleagues.

Secondary prevention can focus on developing self-awareness and providing individuals with a number of basic relaxation techniques. Health promotion activities and lifestyle modification programs also fall into the category of secondary level interventions. They are particularly useful in helping individuals deal with stressors inherent in the work environment that cannot be changed and have to be "lived with," like, for example, job insecurity. Such training can also prove helpful to individuals in dealing with stress in other aspects of their life, that is, non-work related. However, the role of secondary prevention is essentially one of damage limitation, often addressing the consequences rather than the sources of stress which may be inherent in the organization's structure or culture. They are concerned with improving the "adaptability" of the individual to the environment.

Overall, the evidence suggests that such secondary level interventions can temporarily reduce experienced stress (Murphy 1988, 1996). Some recent interventions have found a modest improvement in self-reported symptoms and psychological indices of strain (Reynolds *et al.* 1992; Sallis *et al.* 1984), but little or no change in job satisfaction, work stress, or blood pressure (Murphy 1996). Participants in one company-wide program, for example, reported improvements in health in the short term (i.e. three months post-intervention), but little was known about its long-term effect (Teasdale 1996). Similarly, counseling appears to be successful in treating employees suffering from stress. However, as they are likely to re-enter the same work environment as dissatisfied in their job and no more committed to the organization than they were before, potential productivity gains may not be maximized.

The evidence concerning the impact of health promotion activities has reached similar conclusions. Lifestyle and health promotion activities appear to be effective in reducing anxiety, depression, and psychosomatic distress, but do not necessarily moderate the stressor–strain linkage. According to Ivancevich and Matteson (1988), 70 percent of individuals who attend such programs revert to their previous lifestyle habits after a few years. Furthermore, smoking, alcohol abuse, obesity and coronary heart disease are more prevalent among the lower socioeconomic groups. Members of these groups are likely to occupy positions within the organizational structure which they perceive afford them little or no opportunity to change or modify their job stresses. The health of the "most at risk" individuals is not addressed.

Tertiary prevention

Tertiary prevention is concerned with the treatment, rehabilitation and recovery process of those individuals who have suffered or are suffering from serious ill health as a result of stress. Interventions at the tertiary level typically involve the provision of counseling services for employee problems in the work or personal domain. Such services are either provided by in-house counselors or outside agencies in the form of an employee assistance program (EAP). EAPs provide counseling, information and/or referral to appropriate counseling treatment and support services.

There is evidence to suggest that EAPs and counseling are effective in improving the psychological well-being of employees and has considerable cost benefits (Cooper and Sadri 1991). Based on reports published in the US, figures typically show savings to investment rates of anywhere from 3:1 to 15:1 (Cooper and Cartwright 1994). Such reports have not been without criticism, particularly as schemes are increasingly being evaluated by the "managed care" companies responsible for their implementation and who frequently are under contract to deliver a preset dollar saving (Smith and Mahoney 1989). Like stress management programs, counseling services can be particularly effective in helping employees deal with non-work-related stress (i.e. bereavement, marital breakdown, etc.) and with workplace stressors that cannot be changed but nevertheless tend to spill over into work life.

Tertiary prevention may often therefore be easier than primary or secondary prevention, but it may only be an effective short-term strategy. In focusing at the outcome or "rear end" of the stress process (i.e. poor mental and physical health) and taking remedial action to redress that situation, the approach is essentially reactive and recuperative rather than proactive and preventative (Cartwright et al. 1995).

Action to reduce stress at work is usually prompted by some organizational problem or crisis, for example escalating rates of sickness absence or labor turnover. Consequently, actions tend to be driven by a desire to reduce or arrest costs (i.e. problem-driven-negative motives) rather than the desire to maximize potential and improve competitive edge (i.e. gains-driven-positive motives). The danger of this type of approach is that once sickness absence or labor turnover rates stabilize at an acceptable level, interventions may lose their impetus and be considered no longer necessary. Organizations need to consider stress prevention not only as a means of cost reduction or containment, but also as a means of maintaining and improving organizational health and increasing productivity. The costs of stress and the collective health and wealth of organizations and their workers is of great importance to society as a whole. Occupational health and stress is not just an organizational problem but a wider societal problem, both directly and indirectly through increased taxation and state health insurance contributions or diminished living standards as a result of loss of competitive edge. This final section is therefore concerned with the extent to which economic incentives can be used to address the problem of work stress.

Economic incentives

Typically, organizations respond to statutory legislation by implementing the minimum requirements to conform to the law. Rather than merely punishing "bad practice," the more effective way of encouraging "good practice" is to reward it. This could take the form of providing tax incentives for validated health and safety expenditure incurred by organizations as discussed in the recent European Founding publication (Bailey *et al.* 1994).

Another option is to more directly link risk assessment and stress prevention strategies to insurance premiums. Currently, the cost of employee accidents and compensation for injuries and illness and negligence is met by a variety of insurance bodies in both public and private sectors. Insurance premiums may be levied as a flat rate or vary according to the claims experience of the industry sector or the individual organizations. When premiums are linked to the claims experience or past accident history of the individual organization, employers become more aware of the true cost of their actions. If an employer is penalized by an increased premium as a result of a high accident rate, they are likely to take steps to address and improve the situation. However, there are drawbacks to such arrangements. For example, employers may put pressure on employees not to pursue claims or report accidents. Claims experience databased costs can give a distorted picture when there is a large payment made for a long-term disability or fatality. Similarly, experienced based solely on accident frequency rates may unfairly penalize an organization which has a lot of relatively inexpensive minor accidents compared to an organization with fewer accidents, but which result in a more severe and costly outcome. Most importantly, experienced-based insurance ratings focus on historical record and so do not take into account the efforts an organization may be making to reduce future risk. However, there would perhaps be some benefit in insurance providers pooling their collective experiences and statistics on an industry basis to help to identify particular business sectors which might benefit from more specifically targeted health and safety initiatives.

A more effective and fairer way in which organizations could be rewarded for the efforts in creating more healthy working environments would be to link incentives to stress audits and the presence of stress intervention programs. A rather similar scheme, the Work Injury Reduction Program (WIRP), is currently being trialed in Alberta, Canada. Employers who have voluntarily opted to join the scheme are required to undergo an annual audit of their management systems. This audit focuses on six areas: corporate leadership, operations, human resources, facilities and services, administration and health, and safety information and promotion. The organizations' performance is scored out of a possible 2000 points to provide an index of progress. Employers are required to take action on the results of this audit and the report recommendations to qualify for financial incentives. The potential exists for large companies to receive incentives as high as $2 million.

Conclusion

While there is considerable research activity at the secondary and tertiary level, primary stress reduction strategies are comparatively rare (Cooper and Williams 1997; Murphy 1984). This is particularly the case in the US and the UK. Organizations tend to prefer secondary and tertiary level interventions for several reasons. First, there are relatively more published data available on the cost benefit analysis of secondary and tertiary programs, particularly EAPs (Berridge *et al.* 1992). Second, the professionals traditionally responsible for initiating interventions (i.e. counselors, physicians and clinicians) feel more comfortable with changing individuals than changing organizations (Ivancevich *et al.* 1990). Third, it is easier and less disruptive to change the individual than to embark on an extensive and potentially expensive organizational development program – the outcome of which may be uncertain (Cooper and Cartwright 1994). Finally, secondary and tertiary interventions present a high profile means by which organizations can "be seen to be doing something about stress" and taking reasonable precautions to safeguard employee health.

However, the creation of healthy and productive work organizations also requires attention to job and organizational characteristics, which are causing stress for workers. Until the sources of stress at work are addressed and controlled, either through job redesign or organizational change, it seems unlikely that organizations will be successful in efforts to foster both worker well-being and organizational performance.

References

Bailey, S., Jorgensen, K., Kruger, W. and Litske, H. (1994) *Economic Incentives to Improve the Working Environment*, Dublin: European Foundation for the Improvement of Living and Working Conditions.

Berridge, J., Cooper, C. L. and Highley, C. (1992) *Employee Assistance Programmes and Workplace Counselling*, Chichester: John Wiley.

Cartwright, S., Cooper, C. L. and Murphy, L. R. (1995) Diagnosing a healthy organization: a proactive approach to stress in the workplace, in L. R. Murphy, J. J. Hurrell, S. L. Sauter and G. P. Keita (eds) *Job Stress Intervention: Current Practice and Future Directions*, Washington, DC: American Psychological Association.

Cooper, C. L. and Cartwright, S. (1994) Healthy mind, healthy organization – a proactive approach to occupational stress, *Human Relations* 47(4): 455–71.

Cooper, C. L. and Cartwright, S. (1997) An intervention strategy for workplace stress, *Journal of Psychosomatic Research* 43(1): 7–16.

Cooper, C. L. and Marshall, J. (1976) Occupational sources of stress: A review of the literature relating to coronary heart disease and mental ill health, *Journal of Occupational Psychology* 49: 11–28.

Cooper, C. L. and Sadri, G. (1991) The impact of stress counselling at work, in P. L. Perrewe (ed.) *Handbook of Job Stress* (special issue), *Journal of Social Behavior Personality* 6(7): 411–23.

Cooper, C. L. and Williams, S. (1997) *Creating Health Work Organizations*, Chichester: John Wiley.

Cox, T. (1978) *Stress*, London: Macmillan.

Edwards, J. R. and Cooper, C. L. (1990) The person–environment fit approach to stress: recurring problems and some suggested solutions, *Journal of Organizational Behavior* 11: 293–307.

Elkin, A. J. and Rosch, P. J. (1990) Promoting mental health at the workplace: the prevention side of stress management, *Occupational Medicine State of the Art Reviews* 5(4): 739–54.

Heaney, C. A., Israel, B. A., Schurman, S. J., Baker, E. A., House, J. S. and Hugentobler, M. (1993) Industrial relations, worksite stress reduction, and employee well-being: A participatory action research investigation, *Journal of Organizational Behavior* 14: 495–510.

Holt, R. R. (1982) Occupational stress, in L. Goldberger and S. Breznitz (eds) *Handbook of Stress*, New York: The Free Press.

Ivancevich, J. M. and Matteson, M. T. (1988) Promoting the individual's health and well-being, in C. L. Cooper and R. Payne (eds) *Causes, Coping and Consequences of Stress at Work*, Chichester: John Wiley.

Ivancevich, J. M., Matteson, M. T., Freedman, S. M. and Phillips, J. S. (1990) Worksite stress management interventions, *American Psychologist* 45: 252–61.

Jackson, S. E. (1983) Participation in decision making as a strategy for reducing job related strain, *Journal of Applied Psychology* 68: 3–19.

Murphy, L. R. (1984) Occupational stress management: a review and appraisal, *Journal of Occupational Psychology* 57: 1–15.

Murphy, L. R. (1988) Workplace interventions for stress reduction and prevention, in C. L. Cooper and R. Payne (eds) *Causes, Coping and Consequences of Stress at Work*, Chichester: John Wiley.

Murphy, L. R. (1996) Stress management in work settings: A critical review of the research literature, *American Journal of Health Promotion* 11: 112–35.

Parkes, K. and Sparkes, T. J. (1998) *Organizational Interventions to Reduce Work Stress. Are They Effective? A Review of the Literature*, United Kingdom: Health and Safety Executive, RR193/98, ISBN 0-7176-1625-8.

Payne, R., Wall, T., Borrill, C. and Carter, A. (1999) Stress as a moderator of the relationship between work characteristics and work attitudes, *Journal of Occupational Health Psychology* 4: 3–14.

Quick, J. C. (1979) Dyadic goal setting and role stress in field study, *Academy of Management Journal* 22: 241–52.

Reynolds, S., Taylor, E. and Shapiro, D. A. (1993) Session impact in stress management training, *Journal of Occupational and Organizational Psychology* 66: 99–113.

Sallis, J. F., Trevorrow, T. R., Johnson, C. C., Howell, M. F. and Kaplan, R. M. (1984) Worksite stress management: a comparison of programmes, *Psychological Health* 1: 237–55.

Sauter, S., Murphy, L. R. and Hurrell, J. J. (1990) A national strategy for the prevention of work related psychological disorders, *American Psychologist* 45: 1146–58.

Schaubroeck, J., Ganster, D. C., Sime, W. and Dittman, D. (1993) A field experiment testing supervisory role clarification, *Personnel Psychology* 46: 1–25.

Smith, D. and Mahoney, J. (1989) McDonnell Douglas Corporation's EAP produces hard data, *The Almacan* 18–26.

Teasdale, E. (1996) Stress management within the pharmaceutical industry, in *Stress Prevention in the Workplace*, Dublin: European Foundation for the Improvement of Living and Working Conditions.

Wall, T. D. and Clegg, C. W. (1981) A longitudinal study of group work redesign, *Journal of Occupational Behavior* 2: 31–49.

Index

Note: page numbers in italics and bold refer to figures and tables respectively.